Neutron Scattering in Chemistry

G.E. BACON, MA, ScD (Cantab), PhD, FInstP
Professor of Physics,
University of Sheffield

BUTTERWORTHS
LONDON - BOSTON
Sydney - Wellington - Durban - Toronto

THE BUTTERWORTH GROUP

UK
Butterworth & Co (Publishers) Ltd
London: 88 Kingsway, WC2B 6AB

AUSTRALIA
Butterworths Pty Ltd
Sydney: 586 Pacific Highway, Chatswood, NSW 2067
Also at Melbourne, Brisbane, Adelaide and Perth

SOUTH AFRICA
Butterworth & Co (South Africa) (Pty) Ltd
Durban: 152–154 Gale Street

NEW ZEALAND
Butterworths of New Zealand Ltd
Wellington: 26–28 Waring Taylor Street, 1

CANADA
Butterworth & Co (Canada) Ltd
Toronto: 2265 Midland Avenue,
Scarborough, Ontario, M1P 4S1

USA
Butterworths (Publishers) Inc
Boston: 19 Cummings Park, Woburn, Mass. 01801

All rights reserved. No part of this publication may be reproduced or transmitted in any form or by any means, including photocopying and recording, without the written permission of the copyright holder, application for which should be addressed to the publisher. Such written permission must also be obtained before any part of this publication is stored in a retrieval system of any nature.

This book is sold subject to the Standard Conditions of Sale of Net Books and may not be re-sold in the UK below the net price given by the Publishers in their current price list.

First published 1977

ISBN 0 408 70800 X

© Butterworth & Co (Publishers) Ltd 1977

LIBRARY OF CONGRESS CATALOGING IN PUBLICATION DATA

Bacon, George Edward, 1917–
   Neutron scattering in chemistry.

   Bibliography: p.181
   Includes index.
   1. Atomic theory. 2. Neutrons–Scattering.
I. Title
QD461.B22   541'.28   76–49474
ISBN 0-408-70800-X

Printed and bound in Great Britain by Butler and Tanner Ltd., Frome, Somerset

PREFACE

The use of beams of neutrons to study solids emerged in 1945 as one of the first by-products of the early 'atomic piles', the forerunners of present-day nuclear reactors. Since then the scope and power of neutron scattering techniques have steadily increased but it is only in the last ten years that they have been assimilated as a standard tool of investigation in chemistry, rather than being mainly a preserve of physicists.

The aim of this book is to give an account of those areas in chemistry where neutron techniques make well-established contributions and of the newer areas in which recent work promises very significant advances. At the same time a general account is given of the principles of neutron scattering and some details of the experimental techniques, so that the reader can form some judgement of experimental work in this field from a practical point of view. The author indeed hopes that increasing numbers of chemists will be attracted to enter this field which can contribute to the solution of so many problems of the solid and liquid states, particularly when used to complement results from other techniques such as X-ray diffraction, infrared and Raman spectroscopy and nuclear magnetic resonance.

The author is particularly grateful to the Institutions, editors and individuals who have permitted illustrations from their periodicals and papers to be reproduced, a courtesy without which such a book cannot be written.

G.E.B.

CONTENTS

1	INTRODUCTION	1
2	PRINCIPLES OF NEUTRON SCATTERING	6
	2.1 Nuclear scattering	6
	2.2 Magnetic scattering	11
	2.3 Polarization analysis	16
3	EXPERIMENTAL METHODS	19
	3.1 Elastic and inelastic scattering	20
	3.2 Time-of-flight techniques	28
	3.3 'Hot' and 'cold' sources	31
	3.4 Small-angle scattering	31
	3.5 Choice of material: powder-profile refinement	32
4	STRUCTURAL STUDIES	39
	4.1 Sucrose: a typical study	39
	4.2 Hydrogen bonds	41
	4.3 Molecular overcrowding	45
	4.4 Heavy-element compounds	48
	4.5 Some simple molecules	53
	4.6 Amino-acid derivatives	62
5	DIRECT METHODS OF STRUCTURE ANALYSIS	69
6	CORRELATION OF X-RAY AND NEUTRON DATA: X–N SYNTHESES	73
7	STUDIES OF BIOLOGICAL MATERIALS	82
	7.1 Anomalous scattering	83
	7.2 Small-angle scattering	84
8	MEASUREMENTS OF COVALENCY	92
9	DEFECTS AND NON-STOICHIOMETRY	104
10	MOLECULAR SPECTROSCOPY	116
	10.1 Principles of inelastic incoherent scattering	116
	10.2 Typical applications	120
	10.3 Selective deuteration	127
	10.4 Surface chemistry and catalysis	128

	10.4.1 Transition metal complexes	128
	10.4.2 Molecular-sieve zeolites	130
	10.4.3 Water layers in clay minerals	133
	10.4.4 Ionic solutions	134
11	**POLYMERS**	138
12	**LIQUIDS, GLASSES AND GASES**	150
	12.1 Liquids	150
	12.1.1 Binary alloys	156
	12.1.2 Aqueous solutions	158
	12.1.3 Liquid water	162
	12.1.4 Atomic motion	164
	12.2 Glasses	168
	12.3 Gases	175
	APPENDIX 1	177
	APPENDIX 2	179
	APPENDIX 3	180
	BIBLIOGRAPHY	181
	INDEX	183

1

INTRODUCTION

In 1912 von Laue and his collaborators carried out an experiment with a crystal of $CuSO_4 \cdot 5H_2O$ which demonstrated both that X-rays were an electromagnetic radiation and also that solids had a highly regular form, achieved by a three-dimensional stacking of building blocks, or unit cells, on an atomic scale. This discovery was exploited by the Braggs, subsequently becoming Sir William and Sir Lawrence, to develop the technique of X-ray diffraction as the method of investigating in detail the structures of solids on an atomic scale. For over 30 years this technique, the basis of the concept of X-ray crystallography, remained the undisputed tool which the chemist, metallurgist, mineralogist or physicist must employ if he was to understand properly any material with which he dealt.

Meanwhile in 1927 the diffraction of electrons had been demonstrated and an analogous technique of electron diffraction was developed. This has proved to be of immense value in the study of surfaces and thin films, because of the relatively high absorption of electrons, but does not have the versatility and practical simplicity of its X-ray predecessor.

Discovery of the neutron in 1932 led to the demonstration in 1936 that it, too, could be diffracted and the experiment provided a most convincing demonstration of the wave-particle duality of matter. However, as sources of neutrons the radium-beryllium sources of this period were feeble in the extreme, having an output of neutrons about 10^{13} times smaller than for a conventional nuclear power reactor of the second half of the 20th century. Nevertheless the neutron was a very interesting particle to the theoreticians, possessing zero charge but a substantial magnetic moment, and in 1939 Halpern and Johnson[1] wrote what is still the most important paper describing how neutrons would be scattered by magnetic materials. Then, in 1942, the first nuclear reactor appeared, utterly transforming the intensity which could be achieved for neutron beams. Since then successive developments have increased these intensities by a further factor of 10^4 times and have made available to the experimenter the immense possibilities of exploring the atomic architecture of solids and liquids to which the neutron holds the key. It is the purpose of this book to explain what these possibilities are, in so far as they are of interest to chemists.

The basis of techniques which use the diffraction of radiation to study the structure of solids is the regularity of the unit cell. The size and shape of the unit cell can be specified in terms of the lengths of the edges of the cell and the angles between them. In simple substances, such as many metals and ionic solids like NaCl, the cell edges are mutually

2 Introduction

perpendicular and of the order of 5 Å (0.5 nm) in length. On the other hand, the majority of organic chemicals have the symmetry known as monoclinic in which one of the three coordinate axes is perpendicular to the plane containing the other two, but these latter are not at right angles to each other. More generally, the unit cell may be described in terms of one of the seven crystal systems of the crystallographer, ranging in linear dimensions from a few ångstrom units to more than one hundred ångstrom units in biological substances. The size and shape of a unit cell can be determined directly from measurements of the angular positions of the spectra produced by the diffraction of monochromatic radiation; the wavelength of the radiation will be chosen to be comparable with interatomic distances, i.e. of the order of 1 Å. The atomic content of the unit cell, by which we mean the spatial coordinates of the individual atoms within the cell, is the main aim of our present study: it is more difficult to determine than the cell dimensions and can be determined only indirectly. In order to make clear how this determination is made we shall first recall some familiar facts concerning acoustics and musical notes. A given musical note will sound differently when played by different instruments because although in each case the fundamental note is the same, it will be accompanied by different overtones for the various instruments. The pressure wave as a whole can always be analysed into a series of simple harmonic curves, one for the fundamental frequency and one for each of the multiples of this frequency. The process of resolution into these separate terms is called Fourier Analysis and the description of the terms, and their amplitudes, constitutes the Fourier Transform of the original function describing the variation of pressure. An example in which we analyse a general periodic function is shown in *Figure 1.1*.

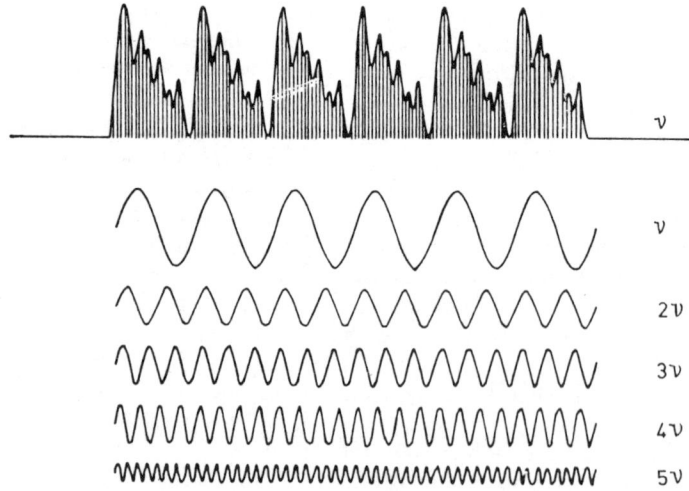

Figure 1.1 The general periodic function of frequency ν at the top of the diagram is compounded by addition of the simple harmonic functions of frequencies ν, 2ν, 3ν, 4ν and 5ν... shown below it

For a solid the existence of the unit-cell provides a periodic function in three dimensions, the function being one which measures the scattering amplitude per unit volume for the radiation which is being used, whether this be electrons, neutrons or X-rays. Each term in the Fourier Analysis is characterized by three indices h, k, l which, in three dimensions, characterize the spatial frequencies of the 'scattering-amplitude waves' along the three coordinate axes, in the same way as the waves of pressure variation for the musical note are characterized by a single spatial parameter. Thus we can describe the scattering amplitude function, as it varies over the unit cell, as a series of terms C_{hkl}.

When radiation is diffracted by the atoms which build up a three-dimensional crystal structure, it appears in those various directions for which it can be considered to undergo Bragg reflection by a set of parallel planes in the crystal, characterized by the so-called Miller indices hkl. The important feature, from our present point of view, is that it can be shown that the amplitude of each hkl diffracted spectrum equals the corresponding Fourier term C_{hkl} which we have just discussed. It follows therefore that if we can measure each of the hkl spectral amplitudes in a diffraction experiment then we can evaluate the scattering amplitude ρ_{xyz}, per unit volume, at any point x, y, z in the unit cell. Thus quantitatively

$$\rho(x,y,z) = \sum_{h,k,l}^{\infty} C_{hkl} \exp 2\pi i(hx/a + ky/b + lz/c) \qquad (1.1)$$

where a, b, c are the edges of the unit cell. This equation is the basis of determination of crystal structure. In assessing its implications we note first of all that it is the variation of *scattering amplitude* over the cell which we are investigating. The nature of this scattering amplitude will depend on the particular radiation which is being used. Thus for X-ray diffraction it is electron density which is explored, because it is the electrons within an atom which are able to scatter the X-rays. For electron diffraction it is electric potential which is important, and for neutron diffraction it is the quantitative ability of various atoms to scatter neutrons. We require to know, therefore, how the ability of atoms to scatter neutrons depends on the nature and constitution of the atom. An important part of our study will be to examine the processes whereby atoms scatter neutrons in order to discover for which atoms location and discrimination will be more effective when neutrons are used than for electrons or X-rays. We shall find that the process of neutron scattering is two-fold, by nuclei and by unpaired, magnetic, electrons and we shall discuss these processes in the following chapter.

Meanwhile we comment that the techniques which employ the scattering of radiation to study the structure of solids, or liquids, extend far beyond the determinations of atomic positions in the unit cell. We emphasize in particular that the distribution of atoms is a changing, dynamic, picture in contrast to a static one. Individual atoms possess thermal energy and are not at rest but in oscillation about their mean equilibrium positions. These motions will not be completely random and uncorrelated, as they would be in a monatomic gas, because there are substantial forces between an atom and its neighbours and displacement of one atom will invariably lead to displacement of others. In simple metals and ionic solids we can regard the individual atomic motions as the superposition of the effects of a wide spectrum of waves of atomic displacement. These waves, which may be either longitudinal or transverse, include the acoustic vibrations of the solid and the quanta of the individual vibrations are designated as phonons. In materials which contain readily identifiable molecules it often pays to look on these as rigid units, or quasi-rigid units, in which the molecule largely maintains its shape but is distorted to some extent by the *intra*molecular stretching, bending or twisting of individual bonds which will produce changes in interatomic separations and angles. At the same time the molecular movements depend on the *inter*molecular forces and these determine the phonon spectrum.

As a result of the thermal motion the atoms will be displaced from their mean positions and this will cause a reduction in the intensities of the diffracted *hkl* spectra; there will be a consequent, but much less noticeable, increase in the diffuse scattering which occurs between the spectra, but which is nevertheless concentrated at the peak or Bragg positions. If the intensity reductions can be measured, and one direct method of doing this is to examine how the intensities fall when the temperature is increased, then the magnitude of the thermal displacements can be deduced. Such deductions are made as part of the standard technique of structural analysis by diffraction methods. There is, however, another way in which the phonons and the intramolecular vibrations of various kinds may be detected and measured. This can be realized by first remarking that the processes of radiation scattering which we have discussed so far are *elastic*. This means that there is no loss of energy in the process of scattering and therefore the scattered beam of, say, neutrons or X-rays will have a wavelength which is exactly the same as that of the incident beam. Nevertheless it is possible for *inelastic* scattering to take place in which the incident neutrons or X-rays lose, or indeed gain, energy by exchange with a quantum of vibrational energy. The latter may excite, or remove, for example, an acoustic phonon or cause the stretching or bending of some interatomic bond within a molecule. The scattered neutrons or X-rays will therefore be of lower or higher energy, i.e. of longer or shorter wavelength respectively in the two circumstances. It turns out that neutrons, but not X-rays, are extremely powerful for investigating these changes, and thus studying vibrational spectra of all kinds. The difference between the two radiations becomes apparent when we consider the energies of a neutron and of an X-ray quantum in relation to the magnitude of a quantum of vibrational energy in a solid. The energy of an acoustic phonon in silicon is of the order of 0.06 eV and a typical energy for the stretching of a hydrogen bond is 0.1 eV. These are of the same order as the energy of a thermal neutron — the energy of a neutron of wavelength 1 Å is 0.08 eV — but are vastly less than the energy of an X-ray quantum. The energy of an X-ray quantum with a wavelength of 1 Å is about 12 keV, i.e. 150 000 times larger than the energy of a neutron of the same wavelength. Thus when an X-ray is inelastically scattered the change of energy or wavelength

is too small to be detected experimentally, but inelastic scattering of a neutron produces changes of energy and wavelength which can be easily and accurately measured. By making a survey of the energy distribution of the inelastically scattered neutrons we can therefore learn about the vibrations of various kinds which occur in the scattering materials.
Several different techniques are available, ranging from a full investigation of single crystals, to determine phonon dispersion curves in defined directions, to a spectroscopic identification of molecular vibrations which can be achieved with powdered or polycrystalline samples.

Reference

1. HALPERN, O. and JOHNSON, M.H., *Phys. Rev.*, **55**, 898 (1939)

Bibliography

For detailed discussion of the main concepts of crystallography and crystal-structure analysis the reader is referred to the four volumes of *The Crystalline State*, Bell, London:
  Vol. I   *A General Survey* by W.L. Bragg (1933)
  Vol. II   *The Optical Principles of the Diffraction of X-rays* by R.W. James (1962)
  Vol. III   *The Determination of Crystal Structures* by H. Lipson and W. Cochran, 3rd edn (1966)
  Vol. IV   *Crystal Structures of Minerals* by W.L. Bragg and G.F. Claringbull (1965)
and to
  An Introduction to Crystallography by F.C. Phillips, Longmans, London, 4th edn (1971)
  An Introduction to X-ray Crystallography by M.M. Woolfson, Cambridge University Press (1970)

2

PRINCIPLES OF NEUTRON SCATTERING

In its passage through a solid or liquid a neutron may be scattered by the atoms in two ways, either by an interaction with the nucleus or by interaction with any electrons of unpaired spin. These unpaired spins are present only in magnetic materials and, accordingly, it is only for magnetic materials that this so-called 'magnetic scattering' can take place. On the other hand, the 'nuclear scattering' takes place, to some degree, for all atoms and we shall consider this first.

2.1 Nuclear scattering

Let us consider a neutron beam, represented by a plane wave, to fall upon a single nucleus fixed at an origin of coordinates. The scattering of the beam will depend on the interaction potential $V(r)$ between the neutron and the atomic nucleus, which are separated by a distance r. In practice the detailed variation with r of this potential is not known, but it is a potential of extremely short range and falls rapidly to zero outside a distance of the order of nuclear dimensions, i.e. beyond about 10^{-12} cm. This latter distance is vastly smaller than the wavelength of the neutrons, which is unlikely to be less than 1 Å (10^{-8} cm) and may be as large as 10 or 20 Å. Consequently the nucleus acts as a point-scatterer, yielding at distance r from the nucleus a scattered beam which can be written as $-b/r$ for an incident beam of unit amplitude. The constant b is called the 'scattering length', or sometimes the scattering amplitude of the nucleus. It has the dimensions of length and is of the order of magnitude 10^{-12} cm. Because, as we have discussed above, the dimensions of the nucleus are very much smaller than the neutron wavelength λ, it follows that the expression $-b/r$ will apply for all values of the angle of scattering. There is no angularly dependent form-factor, as exists, for example, for the X-ray scattering from an atom, where the linear dimensions of the electron cloud in the atoms are about the same as the wavelength of the X-rays.

One of our essential requirements is a knowledge of how the scattering length b varies for different atoms or, more precisely, for different nuclei. For most nuclei it is not possible to calculate b very accurately and it has to be determined empirically by experiment. Values of b for all the elements and for many individual isotopes are given in a table in *Appendix 1* on p. 177 at the end of the book. We emphasize, in considering this table, that b is specific to a given nuclide, i.e. a particular nucleus with a given proton number Z and a given neutron number N. In many cases, for example the metal iron, the values of b are quite different for the different isotopes. We shall understand this more clearly by examining the Breit-Wigner formula for b; this indicates how the value of b depends on two separate contributions, one relating to the physical size of the nucleus and the other depending on the presence of resonant energy-levels within the nucleus. These two contributions are described as 'potential scattering' and 'resonance scattering' respectively and

$$b = R + \frac{\frac{1}{2}\Gamma_n^{(r)}/\kappa}{(E - E_r) + \frac{1}{2}i\Gamma} \qquad (2.1)$$

where R is the nuclear radius, $\Gamma_n^{(r)}$ is the width of the nuclear resonance for re-emission of a neutron and Γ is the total width of the resonance; E is the energy of the incident neutron, E_r is the energy which would give resonance and $\kappa = 2\pi/\lambda$ and is the wave number of the neutron. From the form of equation 2.1 it will be seen that in general b is a complex quantity with real and imaginary parts, but the imaginary component is found to be important only for the few nuclei which have resonances close to thermal energy. Examples are ^{113}Cd and ^{149}Sm, which give rise to anomalous scattering to which we shall refer later because it provides a method of phase determination in structure analysis. The effect of a more remote resonance, which is the more usual circumstance, is to make the value of b either larger or smaller than R, the nuclear radius, and very slightly dependent on the neutron wavelength. This is illustrated in the curve of *Figure 2.1*, which is calculated for a nucleus of radius 10^{-12} cm and a single resonance at 1 eV, which is quite close to the situation for the element rhodium. It must be noted that the scale

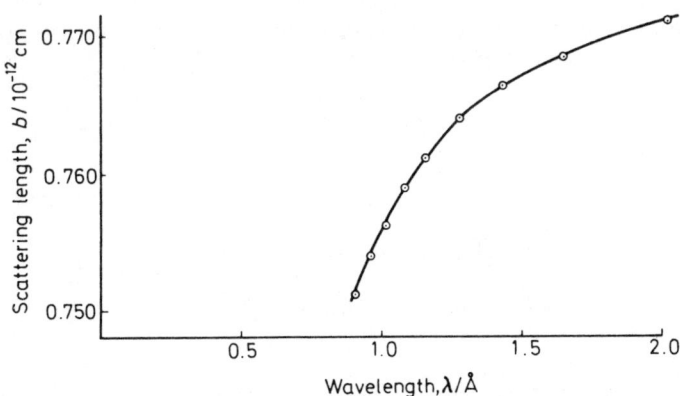

Figure 2.1 The calculated variation of scattering length with neutron wavelength for a nucleus of radius 10^{-12} cm and a single resonance at 1 eV. Note that the scale of ordinates is greatly magnified in order to reveal the small variation with λ

of ordinates in this figure is very expanded and the variations of b with wavelength are really very small.

If there were no effects due to resonance in nuclei then equation 2.1 indicates that b should equal the nuclear radius, which means that, under the prevailing condition that nuclear material is of effectively constant density, b should increase as the cube root of A, which is the mass number of the nucleus and equal to $Z + N$. This general trend through the Periodic Table can be traced in *Appendix 1*, but superposition of effects due to resonance gives at first sight the impression that b varies very haphazardly as we advance from hydrogen to uranium and the trans-uranic elements.

For an element such as nickel, which in its naturally occurring state contains several different isotopes, the effective value of b, which we can distinguish as $\bar{b}$, will equal the average value of b among the isotopes, weighting each isotope according to its relative abundance. Thus

$$\bar{b} = \sum_r w_r b_r$$

where w_r is the fractional abundance of the rth isotope. It is $\bar{b}$ which will determine the intensities of the various hkl reflections in a diffraction pattern. There is, however, another less obvious implication of this 'isotope effect'. In a sample of nickel the different isotopes will be distributed at random among the atomic sites and this leads to a certain amount of disordered scattering which contributes to the background of the diffraction pattern. Rather similar background scattering would appear in the diffraction pattern with either neutrons or X-rays for a disordered alloy AB, where the two components A and B are distributed at random among the atomic sites. The isotope effect with neutrons is called 'isotope incoherence' and can be measured quantitatively in terms of cross-sections*. For a single isotope, or a mono-isotopic element, the scattering length b leads to a single cross-section for scattering σ equal to $4\pi b^2$. When several isotopes are present, the cross-section is subdivided into a coherent part S and an incoherent part s, where

$$\sigma = S + s$$

$$\sigma = 4\pi \sum_r (w_r b_r^2) \qquad (2.2)$$

$$S = 4\pi (\sum_r w_r b_r)^2 \equiv 4\pi \bar{b}^2$$

and

$$s = 4\pi \left\{ \sum_r w_r b_r^2 - (\sum_r w_r b_r)^2 \right\}$$

Other elements which have very pronounced isotope effects for neutrons

*These are commonly expressed in terms of a 'barn', which is a unit of 10^{-24} cm^2.

Principles of Neutron Scattering 9

are hydrogen, titanium and iron. We shall see later that there are many important applications and techniques which depend on the differences between the scattering properties of hydrogen and deuterium.

In considering the resonance scattering contribution for a nuclide it is helpful to think of the *compound nucleus*, formed by a temporary combination of the neutron and the target nucleus, and from which the neutron is subsequently re-emitted. The critical neutron energy E_r is then regarded as the energy which would excite an energy level in this compound nucleus. This concept is particularly useful when we consider the effect of nuclear spin on the scattering of the neutron. The neutron itself possesses a spin of ½, so that if the target nucleus has a spin I then two possible compound nuclei can be formed, having spins of $I + ½$ and $I - ½$, respectively. The energy levels in these two compound nuclei will be different and if resonance effects are significant this will cause a difference between the two values of b. We may denote by b_+ the scattering length when the spins of neutron and nucleus are aligned parallel and by b_- the length when the two spins are antiparallel. In a random neutron-nucleus encounter it is not possible to predict which arrangement of spins will occur, but it is possible to show that on average there is a probability $(I + 1)/(2I + 1)$ of parallel alignment and $I/(2I + 1)$ of an antiparallel arrangement. This behaviour leads to a similar kind of incoherent scattering as we observed for different isotopes and the two probabilities play the same part as did the abundances of the isotopes. This contribution to the background scattering is termed 'spin incoherence'. Certain elements will possess several isotopes, some or all of which have a nuclear spin. In assessing the total amount of incoherent scattering from such an element we must consider in turn each isotope and each possible combination of spin, suitably weighted to take account of abundance and probability respectively when we make the summations required in an extended form of equation 2.2.

We commented earlier that the parallel or antiparallel alignment of spins will take place at random. This is generally the case but not always true, for it is possible to produce beams of polarized neutrons having a defined spin direction and these can fall on a sample containing polarized nuclei, if the polarization is maintained by application of very high magnetic fields at exceedingly low temperatures. Under normal conditions it is possible to use a very sophisticated technique of polarization analysis to assess whether the neutron spins have been turned over in the scattering process and from such observations it is possible to avoid some of the practical difficulties usually posed by the need to distinguish between the concurrent coherent and incoherent scattering. However, in ordinary scattering experiments spin incoherence presents a troublesome problem and this is particularly so in the case of ordinary hydrogen, where it turns out that the incoherent component may be 40 times as large as the coherent contribution. We shall examine in detail this problem, and the quite different situation for deuterium, because of its wide importance in the practical investigation of hydrogenous materials.

For hydrogen the nuclear spin, i.e. the spin of a proton, is ½ and it can be shown that the two scattering lengths b_+ and b_-, as measured most recently by **Koester** and **Nistler**[1], are

$$b_+ = 1.08 \times 10^{-12} \text{ cm and } b_- = -4.74 \times 10^{-12} \text{ cm}*$$

The weighting factors for the two combinations are ¾, ¼, respectively, so that from equation 2.2 we have

$$\sigma = 4\pi \left[\tfrac{3}{4}(1.08)^2 + \tfrac{1}{4}(4.74)^2 \right] \times 10^{-24} = 81.6 \times 10^{-24} \text{ cm}^2$$
$$\text{(barns)}$$

$$S = 4\pi \left[\tfrac{3}{4}(1.08) + \tfrac{1}{4}(-4.74) \right]^2 \times 10^{-24} = 1.8 \times 10^{-24} \text{ cm}^2$$

$$b = \tfrac{3}{4}(1.08) - \tfrac{1}{4}(4.74) \times 10^{-12} = -0.375 \times 10^{-12} \text{ cm}$$

$$s = (81.6 - 1.8) \times 10^{-24} = 79.8 \times 10^{-24} \text{ cm}^2$$

(2.3)

Thus the incoherent scattering which is contributed to the background is about 40 times as large as S which determines the intensity of the Bragg peaks in diffraction patterns. In powder patterns, as distinct from single-crystal patterns, this large background scattering is very troublesome and has to be avoided by recourse to deuterated materials. The background is relatively very large for powdered samples because the incoherent scattering, being isotropic, is contributed by the *whole* of the sample, whereas the diffraction peaks are produced only by a few, correctly oriented, grains of powder. On the other hand, with a single crystal the whole of the crystal also contributes to the peaks. In fact the incoherent scattering is rather less than the value just calculated would suggest, because the values of b_+ and b_- quoted above refer to bound atoms whereas in a crystal an atom is not completely bound but able to recoil to some extent, passing on energy to vibrational modes of the crystal. For a completely free atom the cross-section is reduced from the bound value by the factor $(M/(M+m))^2$, where m and M are the masses of the neutron and nucleus, respectively. For a hydrogen atom this factor is ¼, but for other atoms it rapidly approaches unity as M increases when we move up through the Periodic Table. In practice, in hydrogenous materials the hydrogen atoms behave as though they were partially bound and the actual total and incoherent cross-sections depend on the neutron wavelength, approaching 80×10^{-24} cm² (barns) at long wavelengths (Melkonian[2]). At a typical wavelength of just over 1 Å used in neutron diffraction the experimentally determined cross-section is about 36 barns and this value may be taken as indicating the incoherent contribution to the background scattering. It is emphasized, however, that the *coherent* cross-section and amplitude, S and b, are correctly calculated above since

---

*The occurrence here of a negative sign for the scattering length indicates a change of phase of π in the scattering process compared with what happens for most nuclei: inspection of *Appendix 1* will show that this occurs for a few other nuclei and elements, e.g. ^{62}Ni, Ti, Mn.

they do depend on the *bound* scattering amplitudes: any recoil effects would destroy the coherence of the radiation scattered from neighbouring atoms.

In contrast to the large incoherent scattering for hydrogen which the expressions at equation 2.3 indicate, the situation for deuterium is quite different. A deuterium nucleus has a spin of 1 and the values of b_+ and b_- are, for a bound atom

$$b_+ = 0.952 \times 10^{-12} \text{ cm} \qquad w_+ = \tfrac{2}{3}$$
$$b_- = 0.097 \times 10^{-12} \text{ cm} \qquad w_- = \tfrac{1}{3}$$

Thus from equation 2.2 we have

$$\begin{aligned}
\sigma &= 4\pi \left[\tfrac{2}{3}(0.952)^2 + \tfrac{1}{3}(0.097)^2 \right] = 7.6 \text{ barns} \\
S &= 4\pi \left[\tfrac{2}{3}(0.952) + \tfrac{1}{3}(0.097) \right]^2 = 5.6 \text{ barns} \\
b &= \tfrac{2}{3}(0.952) + \tfrac{1}{3}(0.097) = 0.667 \times 10^{-12} \text{ cm} \\
s &= 7.6 - 5.6 = 2.0 \text{ barns}
\end{aligned} \qquad (2.4)$$

Thus for deuterium only a small part of the scattering is incoherent and deuteration of hydrogenous materials provides a good solution to the difficulties caused by the high background in the diffraction patterns of powdered samples. A table in *Appendix 2* on p. 179 gives information on other elements and isotopes for which the incoherent scattering is significant; it will be noticed in particular that hydrogen and vanadium are quite exceptional.

The great majority of substances of chemical interest which have been investigated with neutrons, whether from a structural or spectroscopic point of view, have contained hydrogen. Next in importance have been compounds of other light elements, particularly the carbides, nitrides and oxides of elements in the middle and at the top of the Periodic Table, such as the rare-earth elements, uranium and the trans-uranics. In each case the virtue of using neutrons is that the scattering amplitudes of C, N and O are of the same order as those of heavy metals. For example, the scattering amplitude of uranium (0.85×10^{-12} cm) is a little larger than that of carbon (0.66) but less than that of nitrogen (0.94). This, of course, is quite different from the behaviour with X-rays, for which the scattering amplitude is proportional to the atomic number and which for uranium is about 15 times as large as for carbon and 12 times as large as for nitrogen. A further limitation arises with X-rays because the scattering amplitude falls off rapidly as the Bragg angle increases.

2.2 Magnetic scattering

Magnetic scattering is of less chemical interest than the fundamental nuclear scattering which we have just discussed, but does have a quite important application to the study of covalent effects in what are predominantly ionic compounds, such as the oxides and fluorides of the transition elements. Accordingly we shall outline the main factors which determine the magnetic scattering.

Although the neutron possesses no electric charge it does possess a magnetic moment and it is the interaction of this moment with the moment carried by atoms in magnetic materials that gives rise to magnetic scattering. We emphasize that it is the electronic magnetic moments, associated with unpaired spins among the outer electronic cloud, with which we are concerned. From a practical point of view it is important to recognize that the magnetic scattering from, for example, an Mn^{2+} ion in antiferromagnetic MnO or an Fe atom in metallic iron, is of the same order as the nuclear scattering and is susceptible to equally accurate measurement. The way in which the magnetic scattering reveals itself in a diffraction pattern will depend on the magnetic character of the substance being examined, particularly whether it is paramagnetic, ferro-, ferri- or antiferro-magnetic. The behaviour is simplest to describe, and the most informative, in the case of *ordered* magnetic arrangements, which include all but the first class of the four classes of magnetic material which we have just listed.

Each magnetic atom has a 'magnetic scattering length' p which is determined by its magnetic moment in terms of the equation

$$p = \frac{e^2 \gamma}{2mc^2} gJf \tag{2.5}$$

$$= 0.27 \, gJf \times 10^{-12} \, \text{cm}$$

The quantities which determine the numerical factor in this expression are e and m, the charge and mass respectively of the *electron*, c, the velocity of light, and γ, which is the magnetic moment of the neutron expressed in nuclear magnetons. The other quantities are the total quantum number J, the Landé splitting factor g, and f, which is an atomic form-factor which depends on 2θ, the angle of scattering. This equation takes account of the general case where the magnetic moment of the atom is accounted for by both spin and orbital contributions. In many compounds of the iron group of transition elements the orbital momentum is quenched by the crystal field and the magnetic moment is then due solely to the electron spin. Under these circumstances the equation takes the simpler form

$$p = 0.54 \, Sf \times 10^{-12} \, \text{cm} \tag{2.6}$$

where S is the spin quantum number.

The form-factor f arises because the electron cloud, which produces the magnetic scattering, has linear dimensions of about an ångstrom unit, of the same order as the neutron wavelength. As a result of this there will be significant phase differences between the scattered contributions from different parts of the atom, leading to a decrease in the resultant scattering as the angle 2θ increases. Only in the forward direction, where $\theta = 0°$, will full reinforcement occur. Thus $f = 1$ when θ is zero but will have fallen to $f = 0.5$ when $(\sin \theta)/\lambda$ has reached a value of about 0.25 Å^{-1}.

In assessing the magnetic scattering we think of our crystal as being populated by magnetic atoms, each endowed with the appropriate scattering length p. This assembly of magnetic scatterers will have a periodic structure built up of unit cells. *Figure 2.2* shows how for a ferromagnetic material the 'magnetic' unit cell will be identical with the ordinary chemical cell, but this will not usually be so for an *anti*ferromagnetic material and the figure shows an example of an antiferromagnetic material for which the magnetic cell is twice as large as the chemical cell in one direction. In such a case as this, additional reflexions will appear in the diffraction pattern and their indices will involve half-integers, e.g. ½00, if they are indexed according to the *chemical* cell. It may be realized that the details of the diffraction pattern will lead to a knowledge of the magnetic arrangement, which may be extremely complicated in certain materials such as the rare-earth metals. We shall not pursue the interpretation of diffraction patterns any further here but emphasize two fundamental points relating to the interpretation of magnetic scattering and

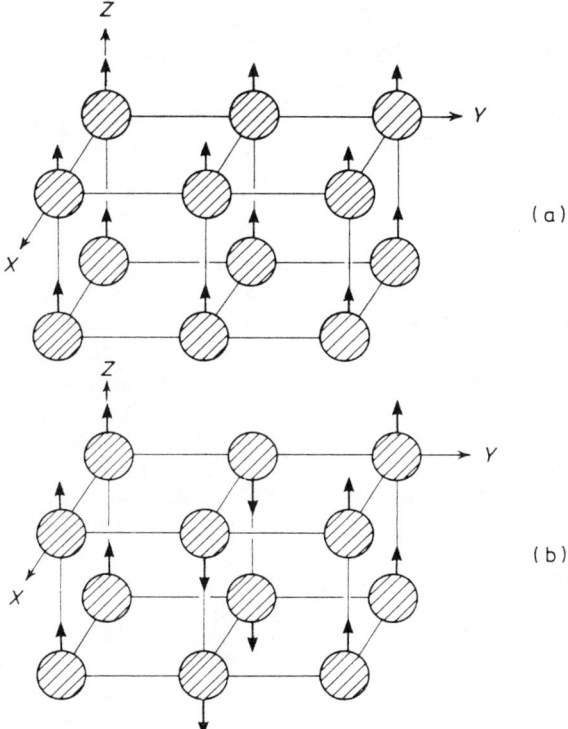

Figure 2.2 For a ferromagnetic material (a) the magnetic unit-cell is of the same size as the chemical cell; for the antiferromagnetic material shown at (b) the magnetic cell is doubled, along the Y axis, compared with the chemical cell

14 *Principles of Neutron Scattering*

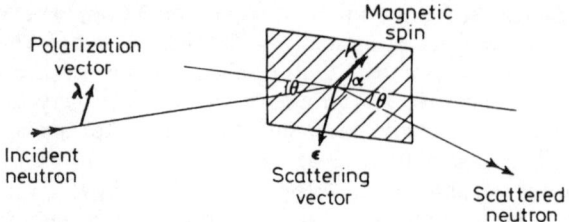

Figure 2.3 Magnetic scattering depends on the vector K parallel to the magnetic spin of the scattering atom in relation to ε, the scattering vector which is perpendicular to the reflection plane. For polarized neutrons the direction of the polarization vector λ is also important

deduction of the underlying magnetic structure. First, we return to equation 2.5 and point out that the amplitude of the neutron beam scattered by an atom depends not only on the magnetic scattering length p but also on the orientation of the magnetic moment vector of the atom with respect to the directions of the incident and scattered beams. The important factor is the angle a, shown in *Figure 2.3*, between the magnetic spin vector K and the scattering vector ϵ, drawn perpendicular to the reflexion plane. The effective scattering amplitude is equal to $p \sin a$, which is zero when the magnetic spin is directed perpendicular to the reflexion plane and has its maximum value, equal to p, when the magnetic spin lies in the reflexion plane so that ϵ and K are perpendicular. More specifically we can say that the effective amplitude is $p|q|$, where the *magnetic interaction vector q* is defined by

$$q = \epsilon(\epsilon \cdot K) - K \tag{2.7}$$

It can readily be demonstrated that the numerical value of q is $\sin a$. This dependence of the scattering on the orientation of K means that from diffraction measurements it is possible to determine the directions of the magnetic moments on the atoms, as well as their locations and magnitudes, thus giving a complete picture of the magnetic architecture. Our second need is to consider the interplay of the nuclear and magnetic scattering and to ask whether these are completely separate and independent of each other or whether, possibly, they are coherent and can produce interference and superposition effects. Most measurements with neutrons are made with unpolarized beams, i.e. with beams in which there are no restrictions on the spin directions of the incident neutrons and these spins can therefore point in any direction. Under these circumstances interference between nuclear and magnetic scattering does not take place and the *intensities*, as distinct from the amplitude vectors, are additive. This means that the intensities of reflexions will depend on terms such as

$$b^2 + |q|^2 p^2 \tag{2.8}$$

where b and p are the nuclear and magnetic scattering lengths which we have defined and q is the magnetic interaction vector already discussed. If, for example, we are studying the diffraction pattern of iron powder (in the absence of any applied magnetic field) then the magnetic spins K may be oriented along any one of the three crystallographic axes, leading to an average value of 2/3 for $|q|^2$ and intensity terms of the form $b^2 + \frac{2}{3}p^2$. For antiferromagnetic materials, as we have seen, there will be certain directions for which magnetic, but not nuclear, reinforcement takes place and here the intensities will depend solely on $|q|^2 p^2$. If, however, the neutron beam is polarized by some device, restricting the incident neutron spins to a single direction, then interference effects between the two types of scattering can be observed. Essentially there is an effective scattering amplitude equal to $b + (\boldsymbol{\lambda} \cdot \boldsymbol{q})$ where $\boldsymbol{\lambda}$ is the polarization vector of the neutron. This behaviour can be shown to lead to a method of actually producing a polarized beam and such a beam is found to have a number of valuable applications. If, as in *Figure 2.4*, a beam of unpolarized neutrons falls upon a ferromagnetic crystal in which all the magnetic spins are aligned upwards by the application of an external magnetic field, then the incident beam is equivalent to two independent beams, one with the neutron spin vector $\boldsymbol{\lambda}$ pointing upwards and the other with downward-spins, as at (*a*) and (*b*), respectively, in the figure. In the two cases the product $\boldsymbol{\lambda} \cdot \boldsymbol{q}$ will equal +1 and −1, respectively, so that the effective amplitudes will be $b + p$ and $b - p$ for the two spin states. By suitable choice of the ferromagnetic material and the reflexion plane it is possible to find circumstances for which $b = p$, and the 220 reflexion of magnetite Fe_3O_4 is such an example. The effective amplitude for downward directed spins will then be zero, so that the reflected beam will consist solely of neutrons whose spin is directed upwards, thus yielding a fully polarized beam.

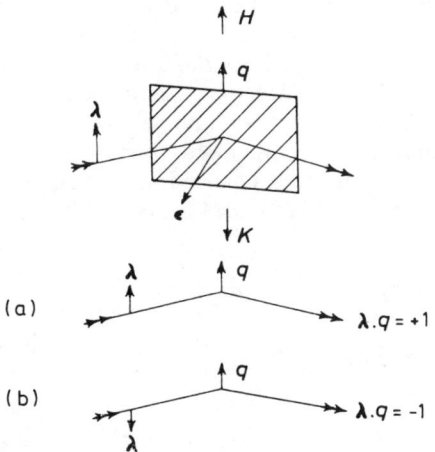

Figure 2.4 Dependence of scattering from a magnetized material, vector K, on the polarization of the incident neutrons, shown for upward- and downward-pointing spins, respectively

2.3 Polarization analysis

Once a polarized beam has been produced it is possible to use it for experiments which observe the changes in neutron polarization which may take place as a result of scattering by an atom. This technique is called polarization analysis (Moon, Riste and Koehler[3]) and leads to direct methods of separating and distinguishing between the different types of scattering which we have previously discussed, e.g. between nuclear and magnetic scattering and between coherent and spin-incoherent scattering. These methods are not yet very widely available, depending as they do on the existence of very intense incident neutron beams which will withstand the relatively inefficient processes of polarization and subsequent analysis, but are likely to become of increasing interest in the future. The essential principles of polarization analysis can be outlined in relation to *Figure 2.5* in which a beam of polarized neutrons of, say, upward-pointing spin is produced from a suitable polarizing crystal. This beam is then diffracted by the sample under study, in a magnetic field, and the diffracted neutrons traverse a system of fields leading to an analyzing crystal which is similar in nature to the polarizer. By operating, or not operating, the spin-flip device along this path it is possible to count, in turn, those neutrons which have, or have not, suffered a spin-reversal during the scattering process. It can be shown that a full description of the scattering process requires the use of Pauli spin matrices (see Halpern and Johnson[4], Marshall and Lovesey[5]) whereby the scattering amplitude is described by

$$U = (u'd')\{b + p\mathbf{q}\cdot\hat{\sigma}\}\begin{pmatrix}u\\d\end{pmatrix} \qquad (2.9)$$

where $\hat{\sigma}$ is the Pauli spin operator and the matrix elements u, d and u', d' refer to the spin states of the incident and scattered beams as 'up' or 'down'. This equation is for a single isotope and a nucleus without spin but can be extended to multiple isotopes and to nuclei of spin I, either unaligned as is normal or maintained in aligned and oriented fashion under magnetic fields at very low temperature. The outcome is that four distinct amplitudes are required to express separately the number of neutrons whose spins have changed, or remained unchanged, in the scattering process. The individual amplitudes in the form given by Moon, Riste and Koehler[3] are

$$\begin{aligned}
\uparrow\uparrow,\ U^{++} &= b - \frac{e^2\gamma}{mc^2}f(S_p)_z + BI_z \\
\downarrow\downarrow,\ U^{--} &= b + \frac{e^2\gamma}{mc^2}f(S_p)_z - BI_z \\
\uparrow\downarrow,\ U^{+-} &= -\frac{e^2\gamma}{mc^2}f\{(S_p)_x + i(S_p)_y\} + B(I_x + iI_y) \\
\downarrow\uparrow,\ U^{-+} &= -\frac{e^2\gamma}{mc^2}f\{(S_p)_x - i(S_p)_y\} + B(I_x - iI_y)
\end{aligned} \qquad (2.10)$$

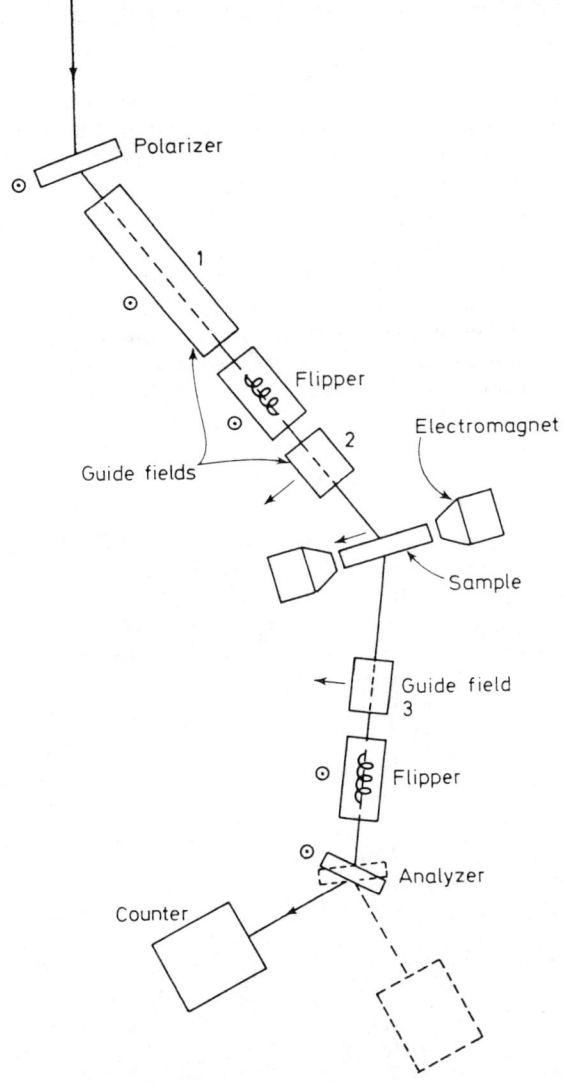

Figure 2.5 The experimental arrangement for polarization analysis. By operating one of the spin-flippers it is possible to distinguish between neutrons whose spin has been changed, or not changed, in the scattering process (From Moon, Riste and Koehler[3], by courtesy of Phys. Rev.)

In these expressions S_p is the projection on the scattering plane of the magnetic spin of a magnetic atom and B is the quantity $b_{\text{incoh}}/\{I(I+1)\}^{1/2}$ where b_{incoh} is the incoherent scattering amplitude. Both S_p and the nuclear spin I are expressed as components along three mutually perpendicular axes x, y, z, of which z is in the direction of the neutron polarization. We mention three conclusions which can be drawn from an examination of these expressions. First, the coherent nuclear scattering, of amplitude b, which gives rise to the Bragg peaks in the diffraction pattern is always non-spin-flip scattering. Secondly, the nuclear-spin incoherent scattering is of two types: the component of spin I_z parallel to the polarization direction produces non-spin-flip scattering, whereas the two perpendicular components I_x and I_y give spin-flip scattering. For the normal case of unaligned nuclei I_x, I_y and I_z are equal so that, in the absence of any magnetic-electron scattering, the non-spin-flip and spin-flip intensities will be proportional, respectively, to $b_{coh}^2 + \frac{1}{3}b_{incoh}^2$ and $\frac{2}{3}b_{incoh}^2$, thus enabling b_{coh} and b_{incoh} to be determined separately. Thirdly, if the neutron polarization direction is along the scattering vector, then $(S_p)_z$ will be zero and, *within the Bragg peaks*, all the spin-flip scattering will be magnetic and all the non-spin-flip scattering will be nuclear; this follows since the incoherent components involving the parameter B in the first two of the equations 2.10 do not contribute to the Bragg peaks.

References

1. KOESTER, L. and NISTLER, W., *Phys. Rev. Lett.*, **27**, 956 (1971)
2. MELKONIAN, E., *Phys. Rev.*, **76**, 1744 (1949)
3. MOON, R.M., RISTE, T. and KOEHLER, W.C., *Phys. Rev.*, **181**, 920 (1969)
4. HALPERN, O. and JOHNSON, M.H., *Phys. Rev.*, **55**, 898 (1939)
5. MARSHALL, W. and LOVESEY, S.W., *Theory of Thermal Neutron Scattering*, Clarendon Press, Oxford (1971)

3
EXPERIMENTAL METHODS

We shall give in this chapter an account of the experimental apparatus and techniques for making measurements with neutron beams, in order to make clear the practical possibilities and also the limitations and difficulties. The first requirement is a beam of neutrons of sufficient intensity and suitable wavelength. The very great majority of measurements, since the first applications in the early 1940s, have used beams of neutrons

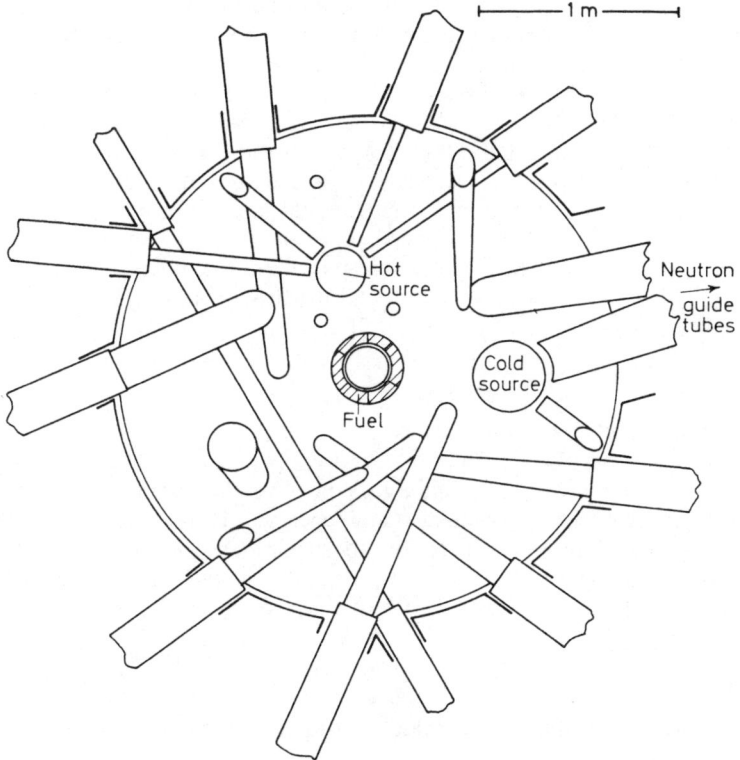

Figure 3.1 A diagrammatic section through the reactor vessel at the Institut Laue-Langevin, Grenoble, showing the way in which the beam-tubes which channel the neutrons to the various spectrometers are grouped around the core of the reactor and its 'hot' and 'cold' sources. The two tubes at the right of the diagram lead to guide tubes which feed the apparatus for small-angle scattering, referred to on p.87 (After Ageron[15], by courtesy of Endeavour)

from nuclear reactors. The scope of the work has increased steadily as technological advances have increased the maximum neutron flux which can be obtained from a reactor. This has increased from about 10^{12} neutrons cm^{-2} s^{-1} in 1945 to 1.2×10^{15} neutrons cm^{-2} s^{-1} in the specially built reactor at the Institut Laue-Langevin at Grenoble which has been in routine use since 1972. Two rather similar reactors are in operation in the USA. We emphasize that these values for the flux of a reactor relate to a position near the core of the reactor reached by the most penetrating beam tubes in *Figure 3.1* and they include neutrons travelling in all directions. This figure, which is a horizontal section through the reactor vessel at the Institut Laue-Langevin, shows how the beam-tubes are grouped around the core of the reactor and its 'hot' and 'cold' sources. For any diffraction or scattering experiment it is necessary to channel neutrons in some kind of collimator in order to produce a beam moving in a defined direction. This means that the number of neutrons falling on a sample placed outside the reactor will be vastly reduced below these values, by a factor of about 10^5. The great majority of reactors used for neutron beam work have been of the continuously operating type, but a notable exception is the pulsed fast reactor at Dubna, USSR, which has been used to produce bursts of thermal neutrons with a half-width in time of 135 μs. For certain techniques a pulsed source is an advantage and it may also well be that further increases in flux beyond 10^{15} will not be forthcoming from a conventional continuous reactor. This has led to the use of electron linear accelerators for producing intense neutron beams in a number of laboratories. With a few exceptions these can still be regarded mainly as exploratory experiments and at the present time the nuclear reactor is still by far the most common source of neutron beams.

3.1 Elastic and inelastic scattering

The changes which a neutron undergoes in a scattering process may involve changes in both momentum and energy. If the incident and scattered neutrons are represented by their wave vectors κ_0 and κ where the magnitude of κ is equal to $2\pi/\lambda$, where λ is the neutron wavelength, then we can define a momentum transfer vector Q such that

$$Q = mv - mv_0$$

where v_0 and v are the incident and scattered velocities. Thus

$$Q = \frac{h}{2\pi}(\kappa - \kappa_0) \qquad (3.1)$$

since $\lambda = h/mv$. We can also specify the energy change as $\hbar\omega$, such that

$$\hbar\omega = \tfrac{1}{2}m(v^2 - v_0^2), \text{ where } \omega \text{ is an angular frequency} \equiv 2\pi\nu$$

$$= \frac{\hbar^2}{2m}(\kappa^2 - \kappa_0^2)$$

$$= E - E_0 \tag{3.2}$$

where E_0 and E are the initial and final energy, respectively.

Our experimental task will be, in general, to make measurements of the changes in κ and E. In the special case of elastic scattering there is no change in E and κ changes in direction but not in magnitude, thus simplifying the measurement and restricting it to a survey of the distribution of neutrons according to direction.

Within a wide range of apparatus which now exists for making experimental observations of neutron scattering, two particular groups may be distinguished. The first type of apparatus is concerned only with the detection of neutrons and measuring their distribution in space, sometimes in three dimensions but often only in a single plane, which is usually the equatorial plane. The second type of apparatus makes specific measurement of the wavelength, in order to determine the energy of the scattered neutrons, and this type will be used for all the measurements of inelastic scattering which are made to ascertain the nature and distribution of vibrations of various kinds, and indeed the details of any interchanges of energy which can take place between a neutron and either

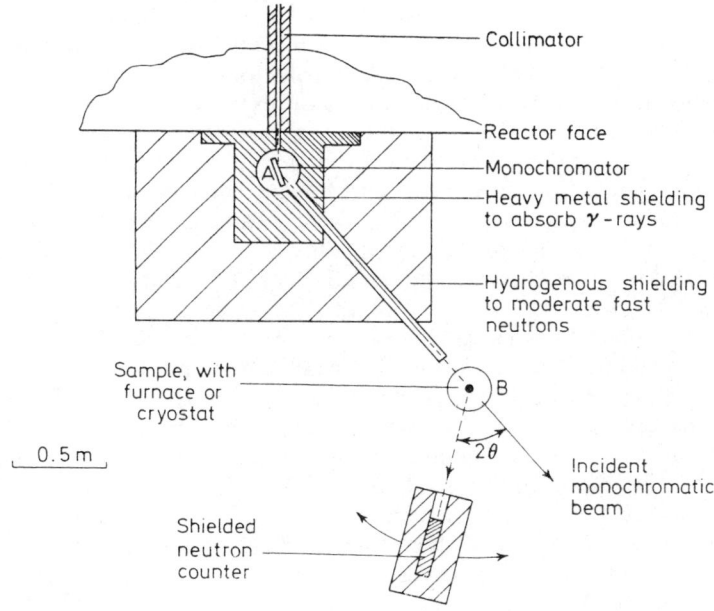

Figure 3.2 The basic details of a two-axis spectrometer for the study of elastic scattering, usually from polycrystalline samples. A monochromatic beam is produced by reflection at the crystal A *and the neutrons subsequently diffracted by a sample at* B *are detected by a counter which rotates about* B.

22 *Experimental Methods*

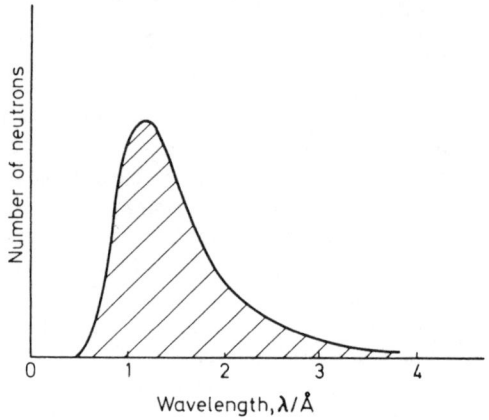

Figure 3.3 The wavelength distribution of the neutrons within a reactor whose moderator is at a temperature of about 40 °C. For the neutrons emerging from a collimator the peak of the distribution is displaced to a slightly longer wavelength

a solid or liquid. Bearing in mind these two distinct categories of measurement, we shall consider briefly the more common types of apparatus.

The most straightforward apparatus for making measurements of elastic scattering is the two-axis spectrometer shown diagrammatically in *Figure 3.2*. The neutrons within the reactor have a wide spectral distribution curve, shown in *Figure 3.3*, and are channelled by a collimator to emerge from the reactor in a single direction, within about ½°. On emergence they fall on a large single-crystal at A, usually of lead, copper or germanium, which selects a narrow wavelength-band from the spectrum. Thus a quasi-monochromatic beam of neutrons of, say, wavelength 1.05 ± 0.02 Å falls on the sample under observation, which is mounted at B on the axis of the instrument. The diffracted neutrons are detected in a counter, most commonly a proportional-counter filled with gaseous $^{10}BF_3$ or sometimes ^3He, which can be rotated around the axis. In this way we obtain the *diffraction pattern* showing the variation of the number of neutrons entering the counter, as a function of 2θ, the angle of scattering. With powdered or polycrystalline samples, for which the individual sets of reflecting planes are oriented randomly in all directions, the material is usually contained in a cylindrical thin-walled can made of vanadium and the measurement is made in the simple manner just described, by either continuous rotation of the counter or, more usually, employing a stepped motion with successive rotations of about 0.1°. For a flux of 2×10^{13} neutrons cm^{-2} s^{-1} a volume of powder of about 1 cm^3 is adequate, with proportionately less for higher fluxes. Vanadium is chosen for the sample holder because, as *Appendix 2* confirms, its scattering is almost entirely incoherent and it does not contribute any spurious peaks to the diffraction pattern of the material which is being studied. *Figure 3.4* shows a typical diffraction pattern for nickel powder.

A more recent technical development at the Institut Laue-Langevin increases the rate of collecting data by providing 400 BF_3 counters,

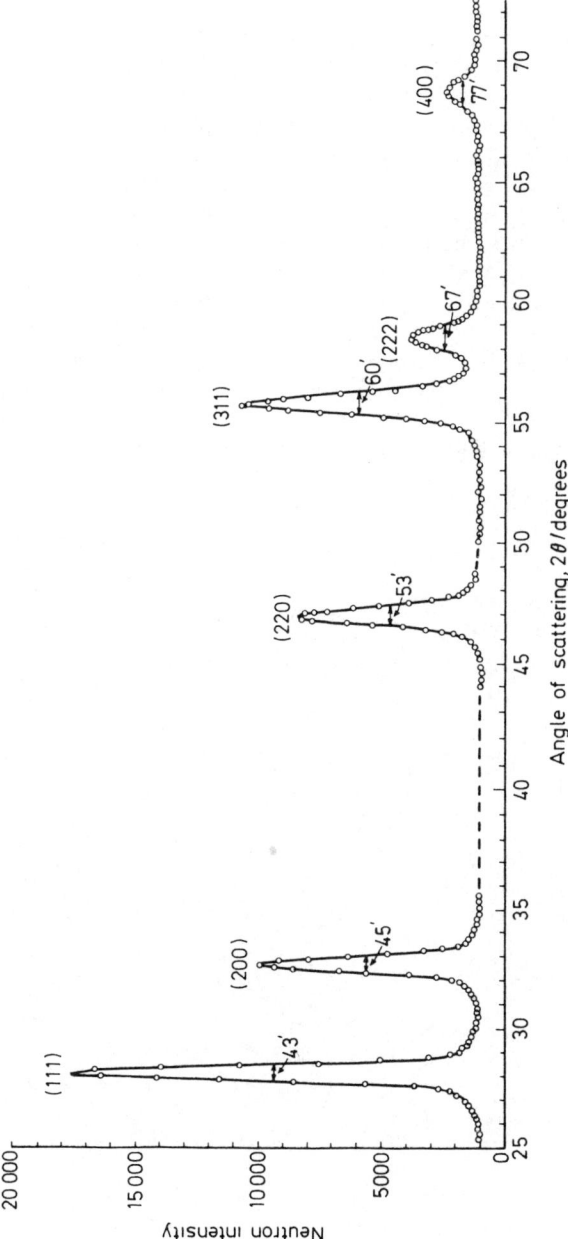

Figure 3.4 A portion of the powder-diffraction pattern for nickel at a wavelength of 1 Å, using a lead crystal as monochromator (From Caglioti and Ricci[14], by courtesy of Nucl. Instrum. Methods)

24 *Experimental Methods*

spaced uniformly around the diffractometer to cover an arc of scattering-angle of $2\theta = 80°$ and giving simultaneous measurement of the neutron intensity at each of these positions. If, on the other hand, a single crystal is being examined (and this will generally result in much more informative conclusions) then the procedure is more complicated. For a given *hkl* plane a reflection will only be obtained when this plane is at the correct inclination θ to the incident beam and the diffracted beam will then appear at the corresponding value of 2θ. It is therefore necessary to orientate the crystal correctly with respect to the counter and to maintain the correct θ, 2θ relationship as the crystal and counter move in synchronism through the reflecting position. This relationship is achieved by suitable linkage between the shaft carrying the crystal and the concentric shaft which carries the counter. In practice the correct orientation and rotations are achieved for each reflecting plane in turn by controlling the movements of the shafts by a pre-programmed punched tape or some similar device. With a single crystal the reflecting planes have to be correctly oriented in three dimensions. The usual procedure is to restrict the counter to movement in the horizontal equatorial plane but to provide rotation about three mutually perpendicular axes for the crystal,

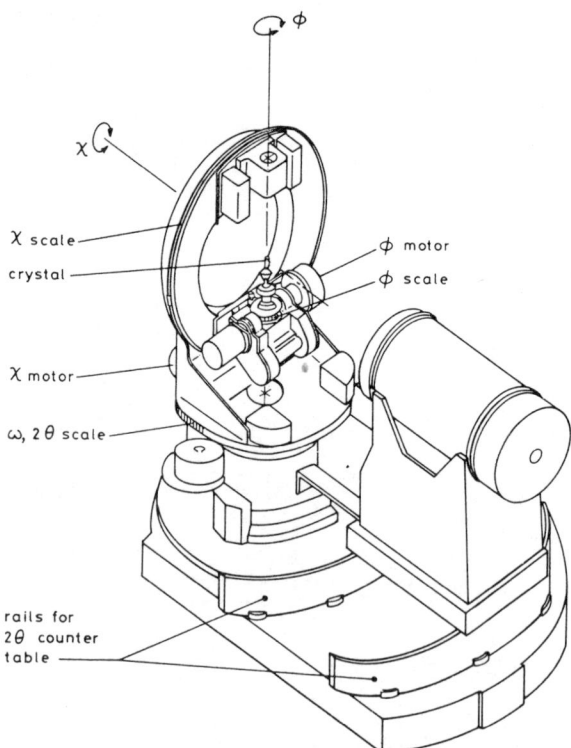

Figure 3.5 *A sketch of a four-circle diffractometer at A.E.R.E., Harwell, used for the recording of single-crystal diffraction patterns. The shielded BF_3 counter is on the right (Reproduced by courtesy of the U.K.A.E.A.)*

thus leading to the so-called 'four-circle diffractometer' which is shown in *Figure 3.5*. A crystal measuring about $(2 \text{ mm})^3$ is required with a flux of 2×10^{13} neutrons cm^{-2} s^{-1}; at a high-flux reactor, $(0.6 \text{ mm})^3$ may be sufficient. It may be commented that these instruments are rather similar in principle to the corresponding instruments used for X-ray diffraction but are considerably larger and heavier. This is mainly due to the fact that substantial amounts of shielding have to be incorporated along the path of the neutron beam and around the detectors, both to reduce the health hazard to personnel and to prevent spurious neutrons from reaching the detectors. Thermal neutrons can readily be absorbed by boron or cadmium but fast neutrons can only be absorbed if they are first slowed down by bulky hydrogenous material such as water, wood or polyethylene.

The most direct, but not always the most convenient or efficient, way of extending the measurements to study inelastic scattering is to add a third axis to the spectrometer of *Figure 3.2*, thus producing what is known as a triple-axis spectrometer, shown diagrammatically in *Figure 3.6*. The third axis, which itself rotates around the axis which carries the sample, carries a further single crystal about which the neutron detector rotates. This crystal and detector enable the wavelength of the scattered neutrons to be measured. For any value 2θ of the scattering angle we are therefore able to perform a wavelength (or energy) analysis of the scattered neutrons. Some typical results for a single crystal of copper by Maliszewski *et al.*[1] are shown in *Figure 3.7* and for the peaks in the curve it is possible to compute Q and ω values for the scattering

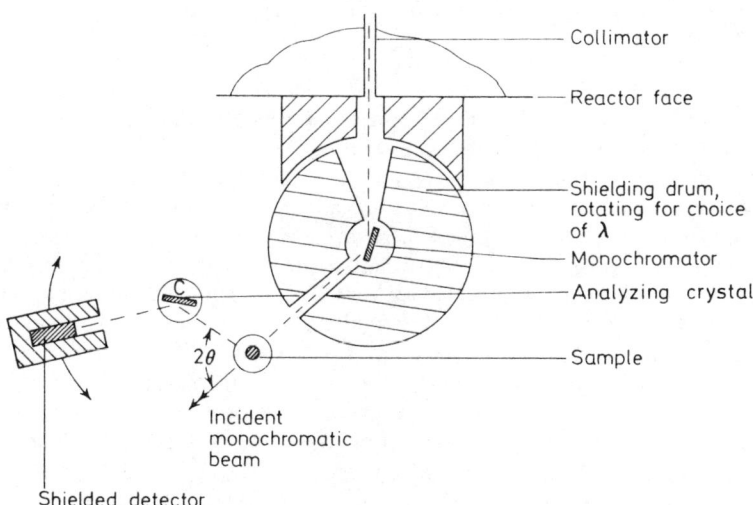

Figure 3.6 Diagram of a triple-axis spectrometer for the energy analysis of inelastically scattered neutrons. The wavelength of the incident neutrons can be varied by rotation of the large shielding-drum; an energy analysis of the neutrons scattered by the sample at angle 2θ is performed by rotation of the analyzing crystal C, with concurrent rotation of the detecting counter at twice the angular velocity

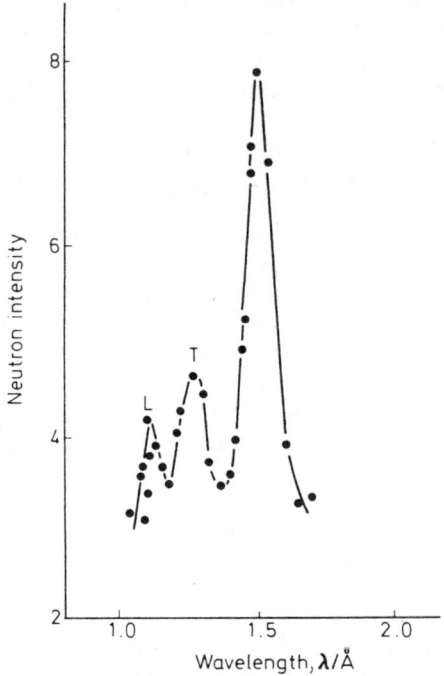

Figure 3.7 The result of a wavelength analysis of the neutrons scattered by a single crystal of copper for an incident wavelength of 1.53 Å. The large peak corresponds to elastic incoherent scattering; the peaks T and L correspond respectively to gain of energy from transverse and longitudinal phonon vibrations (From Maliszewski et al.[1], by courtesy of I.A.E.A.)

process. In practice a more direct procedure is what is called the 'constant Q' method, introduced by Brockhouse[2] and also described by Iyengar[3] in which the values of Q and the phonon momentum q are predetermined and, for a given value of κ_0, a series of satisfactory pairs of values of κ and the crystal-orientation angle are selected by computer. Each pair will correspond to a different value of energy exchange $\hbar\omega$, so that a plot of neutron count against ω will identify the phonon peak. Subsequently phonons of different magnitude may then be explored in turn, for a given direction of travel in the crystal.

The triple-axis spectrometer ensures very accurate and precise analysis, but a number of other methods have been developed in order to increase the rate at which the inelastic scattering data are collected. In the multi-angle reflecting crystal method conventionally known as the MARX spectrometer (Kjems and Reynolds[4]) the analysing crystal and detector of the triple-axis instrument are replaced by a much larger analysing crystal and a linear position-sensitive detector, as shown in *Figure 3.8*. This large crystal accepts neutrons which leave the scattering sample over a wide range of both 2θ and energy. The value of X at the position where the neutron arrives in the detector determines uniquely both $2\theta_s$ and

Experimental Methods 27

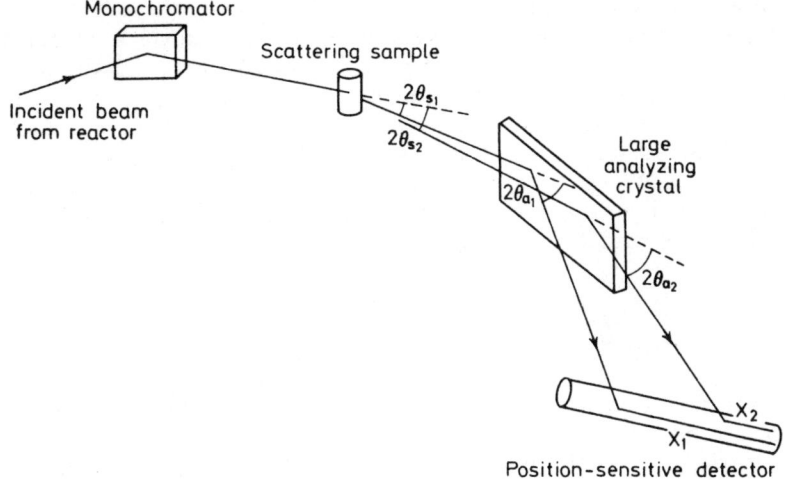

Figure 3.8 Diagram of a multi-angle reflecting crystal (MARX) spectrometer in which the value of X where the neutron enters the position-sensitive detector will determine uniquely both θ_s and θ_a and, therefore, the value of the energy of the neutron

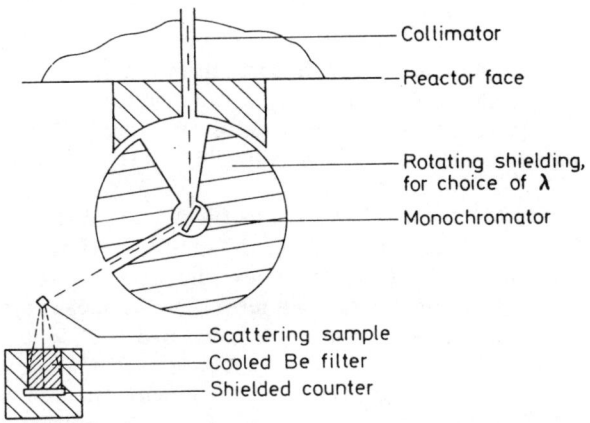

Figure 3.9 A modification of the triple-axis spectrometer (shown in Figure 3.6) for use with a beryllium-filter detector in the energy-loss technique. The incident wavelength is variable but in all cases only neutrons with an energy, after scattering, which is less than 0.005 eV are detected by the counter. Because of the range of scattered angle which is received, the momenta of the neutrons are not exactly known

$2\theta_a$ and, hence, the neutron energy. Thus data for scattered neutrons over a considerable range of Q and ω are collected concurrently. Another modification of the triple-axis instrument uses a beryllium-filter detector (Woods et al.[5]) as shown in Figure 3.9. The only neutrons which are examined are those whose energy after scattering is less than 0.005 eV, i.e. those with wavelengths greater than 4 Å, which is the cut-off wavelength for beryllium. The energy spectrum for the sample is obtained by varying the incident neutron energy, subtracting from it in each case an energy of 0.005 eV to derive the value of the energy transfer. The virtue of this method is that a much larger solid angle can be used for the neutrons leaving the scattering sample. This means that the momenta of the scattered neutrons are known with less certainty, but in many applications this may not be important. At high energies the resolution is much better than for the time-of-flight technique.

3.2 Time-of-flight techniques

The velocities of thermal neutrons are within the range of 10^3–10^5 metres per microsecond and this means that they can readily be measured by time-of-flight techniques using flight paths of the order of a few metres. These techniques offer alternative procedures for both elastic and inelastic scattering investigations. For elastic scattering the method was developed mainly by Buras[6] and his collaborators, using the pulsed reactor at Dubna, for which it is particularly suited. Equally well it can be used with a pulsed source of neutrons produced from a linear accelerator. In either case the diffraction pattern consists of a plot of neutron intensity versus λ, in contrast to a variation with the angle 2θ. The measurement is accordingly made at a fixed angle of scattering and examples are shown in *Figure 3.10*, which gives diffraction patterns for powdered aluminium using respectively a linear accelerator and a pulsed reactor. The method is particularly valuable for studying reflections which have large interplanar spacings. Such reflections select neutrons of long wavelength, since $\lambda = 2d \sin \theta$ and θ is fixed, and both the resolution and intensity of reflection are large under these conditions. On the other hand, when it is important to compare accurately the intensities of individual reflections it is essential to know precisely the wavelength-spectrum of the incident neutrons. Moreover, with single crystals there may be greater uncertainties in the corrections which need to be applied for secondary extinction — the fact that a block within a mosaic crystal may be shielded from the radiation by an identically-oriented block closer to the surface — since these corrections will vary with wavelength.

With a conventional reactor which gives a continuous neutron beam the time-of-flight method can only be used by first pulsing the incident white beam with some form of mechanical chopper. The great majority of measurements of elastic scattering have been made using monochromatic beams and observing the variation of intensity with scattering angle. However, interest in time-of-flight techniques applied to unmonochromated or 'white' beams is important because of the possibility that future increases in the intensity of neutron beams may depend on the use of pulsed sources.

For inelastic scattering, time-of-flight techniques are very widely employed and are in much more general use than the triple-axis spectrometer which

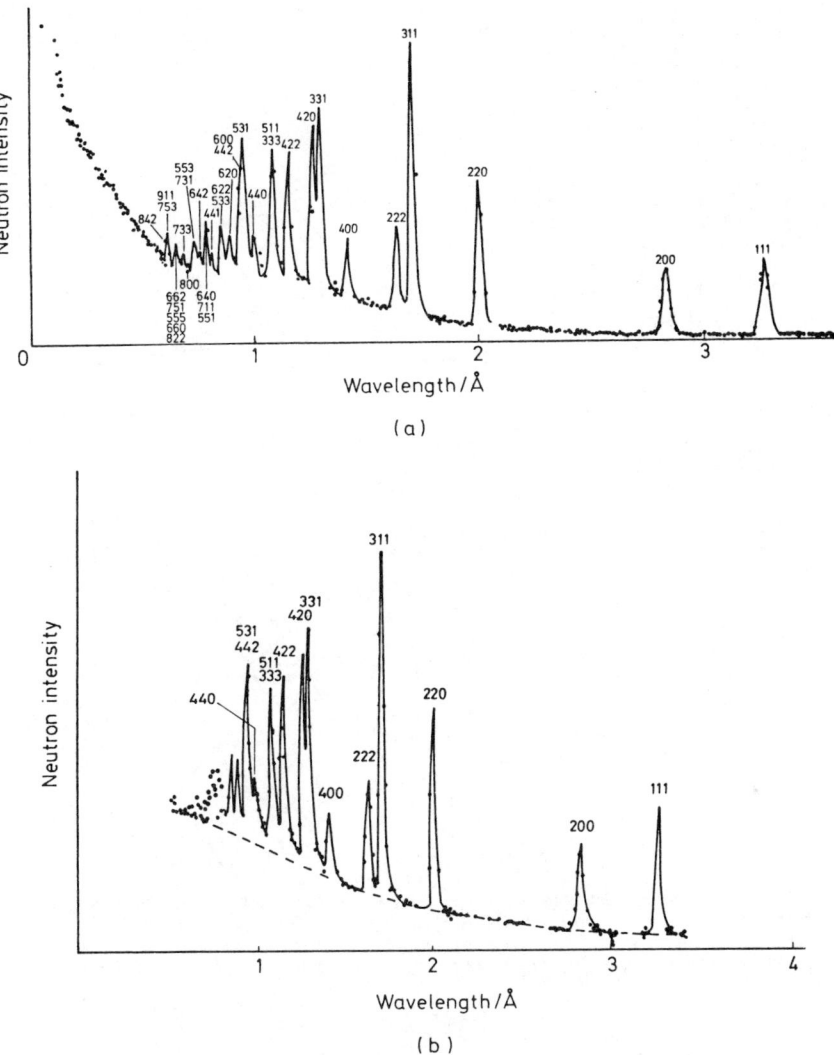

Figure 3.10 Diffraction patterns of aluminium powder recorded by time-of-flight analysis of the scattered neutrons using (a) a linear accelerator (Kimura et al.[16], by courtesy of Nucl. Instrum. Methods) and (b) a pulsed reactor (Buras[6], by courtesy of AEC-ENEA). In each case the measurements were made at a scattering angle $2\theta = 90°$

we have already described. For example, *Figure 3.11* shows a twin-chopper instrument in which the first chopper pulses and roughly monochromates the neutron beam and the second chopper improves the monochromatization. The advantage is that data can be collected at many angles of scattering simultaneously by employing many counters, as shown in the figure. A much simpler form of time-of-flight instrument is shown diagrammatically in *Figure 3.12*. A block of cooled beryllium about 30 cm

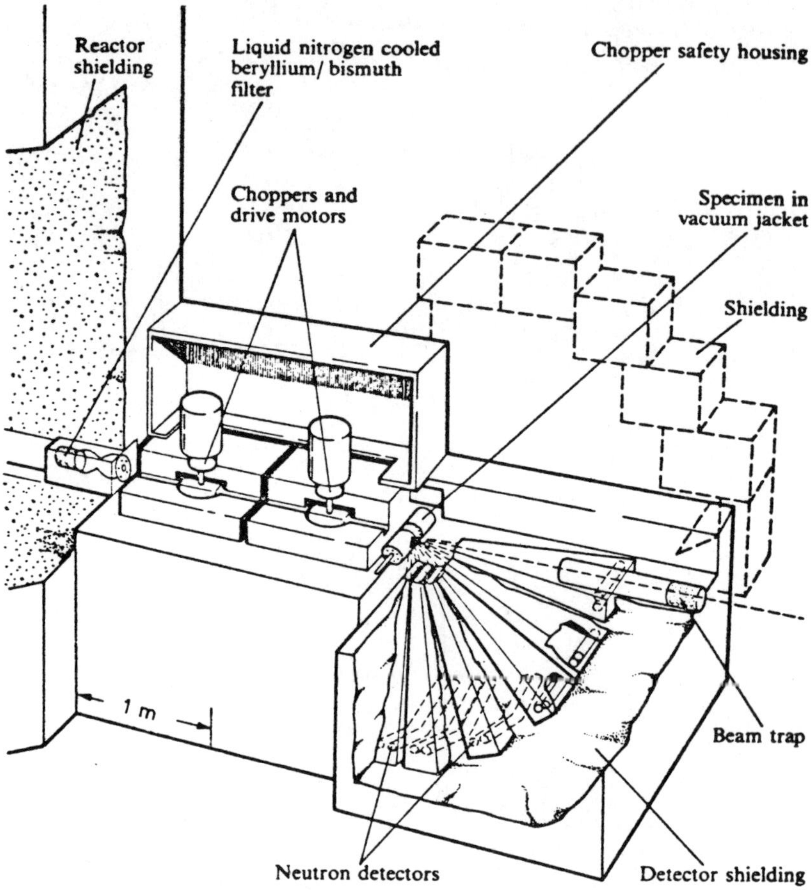

Figure 3.11 Cut-away drawing of a twin-chopper type of time-of-flight spectrometer installed at the DIDO reactor at A.E.R.E., Harwell. The scattered neutrons are recorded simultaneously in the series of counters shown, covering a range of scattering angle, 2θ, from 0 to $90°$ (From Stirling[17], by courtesy of Oxford University Press)

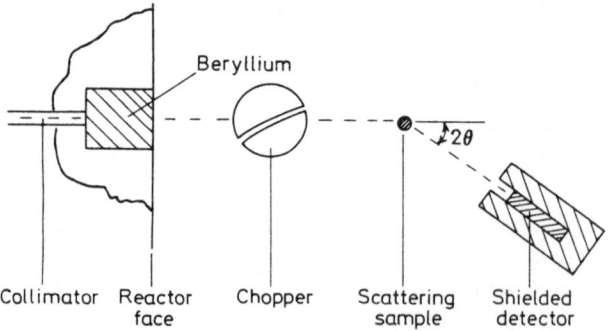

Figure 3.12 Diagram of a simple beryllium-filter spectrometer in which the incident neutrons have been partially monochromatized by passing through a cooled block of beryllium and chopped to permit time-of-flight analysis, employing the energy-gain technique

in length allows only the tail of the neutron spectrum to pass, consisting very predominantly of neutrons with a wavelength just greater than 4 Å, or energies just below 0.005 eV, and this beam is then pulsed by a simple chopper. The neutrons which are observed after scattering will be those which have *gained* energy, since the incident neutrons do not possess sufficient energy to excite vibrations, and the method is accordingly limited by the fact that the higher energy levels in materials will be very little occupied at ordinary temperatures, setting a limit of about 0.080 eV to the energy levels which can be investigated. This situation is to be contrasted with the use of a beryllium filter as a *detector*, which we showed earlier in *Figure 3.9* and for which it is neutrons which have *lost* energy which are being examined. Accordingly the beryllium filter detector can handle a much larger range of energies and is specially useful up to about 0.25 eV.

3.3 'Hot' and 'cold' sources

The wavelength at which the greatest neutron intensity is obtained from a reactor will depend on the temperature of the moderator. For a typical moderator temperature of 40 °C in a heavy-water reactor the peak of the spectrum occurs at a wavelength of about 1.1 Å with the typical spectrum which we showed earlier in *Figure 3.3*. Beyond a wavelength of 3 Å the neutron beam is very weak and this is a great disadvantage for those neutron studies which require long wavelengths. An example of such an application is the study of defects in solids, which, as we shall see later in Chapter 9, depends on using neutrons which have wavelengths which are so long that Bragg reflection cannot take place. This requires that λ should be greater than twice the maximum interplanar spacing of the material, usually meaning that λ is greater than about 5 Å. The output from the reactor of these neutrons of long wavelength can be substantially increased by selectively cooling the portion of the moderator which is acting as the source of neutrons for the particular spectrometer. Conversely, for some purposes it is advantageous to increase the number of short-wavelength neutrons, particularly in the region of about 0.5 Å, and this can be achieved by incorporating a 'hot source' in the reactor. In the high-flux reactor at the Institut Laue-Langevin a volume of 25 litres of liquid deuterium is cooled to 25 K to supply 'cold' neutrons and a graphite block, 10 litres in volume, is heated to 2000 K to supply neutrons of shorter wavelengths. The 'cold' and 'hot' sources have spectral peaks at about 3.5 and 0.5 Å, respectively, and *Figure 3.13* shows the improvement in flux compared with the heavy-water moderator maintained at normal temperature.

3.4 Small-angle scattering

Finally we mention apparatus designed especially for the measurement of small-angle scattering, i.e. the scattering within about a degree of the.

32 *Experimental Methods*

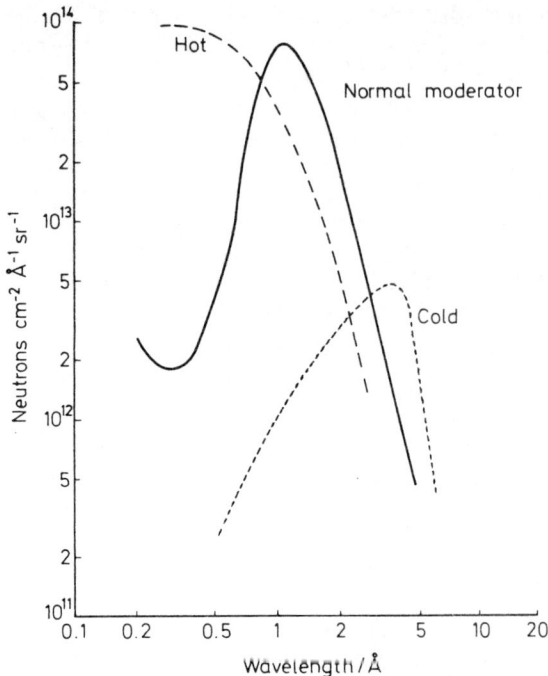

Figure 3.13 Illustration of the principle of 'hot' and 'cold' sources. The curves in the figure show the neutron intensity as a function of wavelength for beam tubes which face, respectively, —— a normal heavy-water moderator at 300 K, - - - a hot source at 2000 K and ······ a cold source at 25 K (Reproduced by courtesy of the Institut Laue-Langevin)

primary beam. By using neutrons of wavelength up to about 20 Å it is possible to examine and measure structural entities ranging in size from 50 to 5000 Å, thus covering such diverse features as clusters in alloys, the conformation of polymer chains and the dimensions and repeat-distances in large biological molecules. The essentials for this kind of measurement are a 'cold source' to give large numbers of neutrons of long wavelength and the use of curved guide-tubes to give highly collimated beams free from unwanted radiation such as X-rays and fast neutrons. The installation at Grenoble has been described by Schmatz *et al.*[7]. In order to secure adequate angular resolution, distances of up to 40 m are used between the sample and the detector.

3.5 Choice of material: powder-profile refinement

In the succeeding chapters of this book we shall consider the application of neutron scattering and neutron diffraction techniques to various kinds of chemical problem. In general we shall be able to see a clear distinction between problems which are essentially structural, requiring knowledge of atomic positions, and those which are essentially dynamical, being concerned with atomic motions and exchanges of energy. In studying the two types of problem we shall be concerned mainly with elastic and inelastic scattering, respectively, and the appropriate measuring apparatus will be used. At the same time we shall see that the results which are forthcoming will depend on the physical nature of the material which is available, not only whether it be solid, liquid or gaseous but, in the former case, whether it be a single-crystal, a powder or polycrystalline block or a glass. In general the information which we can obtain will be most complete and free from possible ambiguities if we can examine single crystals, since these will preserve fully all the directional correlations which are germane to the anisotropic three-dimensional structure. However, single crystals of many materials are extremely difficult to produce in adequate sizes, particularly for those materials which undergo phase changes with change of temperature. Consequently any techniques which are applicable to powdered materials are extremely valuable and we emphasize two of these before proceeding to describe particular applications. First, we shall see that studies of molecular spectroscopy for materials which contain hydrogen rest on assessment of incoherent scattering and these measurements are carried out using powders. Secondly, in suitable cases, structural studies can be made with powdered material by using the technique of profile refinement. This depends not on individual measurements of the intensities of discrete *hkl* reflections but on a comprehensive survey of the *whole* of the diffraction pattern, making individual measurements at intervals of $0.1°$ of the scattering angle 2θ, an interval which is small compared with the angular width of an individual reflection. In this way of the order of 1000 intensity measurements can be made and correlated with calculations made in terms of chosen model structures. The power of the method rests on its value in studying materials for which sufficient information already exists, generally from X-ray studies, to postulate suitable model structures in terms of a limited number of identifiable parameters. In practice the correlation between the measured and calculated values of the ordinates of the diffraction pattern is made by a least-squares analysis, using both structural and instrumental parameters. In favourable cases up to 50 parameters have been assessed.

The applicability of this method of profile refinement, which was originally developed by Rietveld[8], rests on the fact that the diffraction peaks in a neutron powder pattern are almost exactly gaussian in shape, except at very small angles, and their angular width H at angle 2θ can be expressed using three parameters U, V, W as

$$H^2 = U \tan^2 \theta + V \tan \theta + W \tag{3.3}$$

Figure 3.14 illustrates how well a gaussian curve can be fitted to the measured shape of a diffraction peak. Below an angle 2θ of about $30°$ a correction has to be made for asymmetry of the diffraction peaks. The instrumental parameters U, V, W and the neutron wavelength λ are accompanied in the refinement by the usual structural parameters, i.e. a

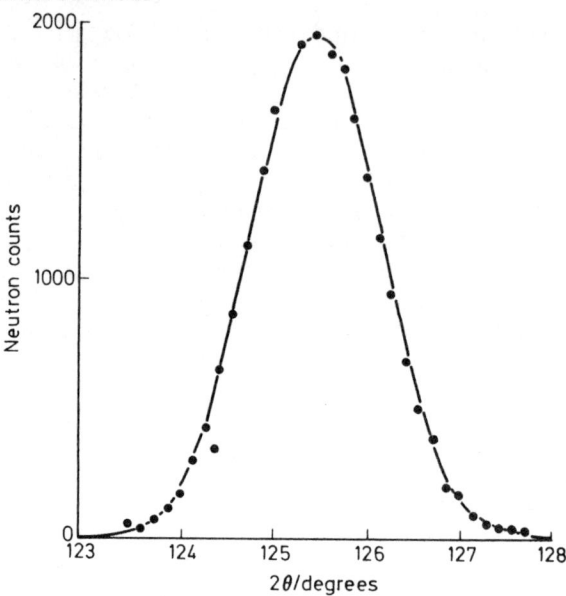

Figure 3.14 A comparison of the shape of an experimentally measured powder-diffraction peak with a Gaussian curve (From Rietveld[8], by courtesy of J. Appl. Cryst.)

scale factor, the dimensions and angles of the unit cell and the coordinates and thermal parameters of the individual atoms. Clearly the accuracy of the resulting analysis will depend on the total number of parameters which have to be determined. We give some examples of this technique to illustrate the type of problem to which it may be applied. Loopstra and Rietveld[9] first applied the method to study a series of metal uranates, neutrons being used in order to give accuracy to the positioning of the oxygen atoms in the presence of the uranium and other heavy atoms. They employed a longer neutron wavelength than normal, using $\lambda = 2.576$ Å, to improve the angular resolution of their patterns. Their results may be summarized by *Table 3.1* which lists, in particular, the number of observations, the number of parameters determined, and the mean standard-deviation of the atomic coordinates. As would be expected, the standard deviation increases with the number of parameters which need to be determined. The precision of the analysis may be expressed in two ways, either by using the normal reliability factor R_B for the integrated intensities of the Bragg reflections or in terms of a profile reliability factor R_P expressed as

$$R_P = 100 \frac{\sum_i (y_{obs} - y_{calc})}{\sum_i y_{obs}} \qquad (3.4)$$

where y_i is the intensity at an angle θ_i in the diffraction pattern and the summation is made over all the measured positions i.

Table 3.1

Compound	Number of ordinates measured	Number of independent reflections	Number of parameters	R value/% Profile R_P	Normal R_B	Standard deviation of coordinates/Å
$CaUO_4$	300	19	13	4.8	1.5	0.0007
$SrUO_4$	913	71	20	11.3	5.8	0.0040
$BaUO_4$	905	76	20	13.0	5.1	0.0052
Ca_2UO_5	1114	160	34	12.6	7.6	0.0084
Sr_2UO_5	1044	160	34	9.8	3.7	0.0074
Ca_3UO_6	1079	115	42	10.8	7.2	0.0238

Hewat[10] has studied a number of ferroelectrics. It may be recalled that the classical ferroelectric KH_2PO_4 has been much studied by neutrons both above and below the ferroelectric change-point, using single crystals, which can readily be grown. On the other hand, the corresponding ammonium salt, $NH_4H_2PO_4$, has proved impossible to produce as sizeable crystals at low temperature and cooling of single crystals produced at room temperature always results in shattering when the phase change takes place.
Hewat has studied the powdered material, albeit in the deuterated form, to determine the details of the hydrogen bond. The importance of deuteration, in order to prevent the counting statistics from being much impaired by the heavy background caused by the incoherent scattering from hydrogen, has to be emphasized.

Jacobson, Collins and Fender[11,12] have examined the perovskite-like compounds $Ba_5(W_3Li_2)O_{15}$, $Ba_4M_3LiO_{12}$ and $Ba_5M_3Li_2O_{15-x}$ where M is tellurium or uranium. In these cases both position parameters and occupation parameters for the sites need to be refined. Around 15 parameters needed to be refined in each case, this number being limited by making the measurements at a temperature of 4.2 K, where thermal motion is very small and a single common temperature-factor could be used for all the atoms.

Perhaps the most striking example of this kind of study so far is provided by Van Dreele and Cheetham's refinement[13] of the structures of $TiNb_2O_7$ and $Ti_2Nb_{10}O_{29}$, where again the advantage of using neutrons is that they define accurately the positions of the oxygen atoms in the presence of the heavy metal atoms. In each case the measurements were made at 4.2 K, to permit the use of a single temperature-factor, and 43 and 56 parameters were refined in the two cases, utilizing 1077 and 1116

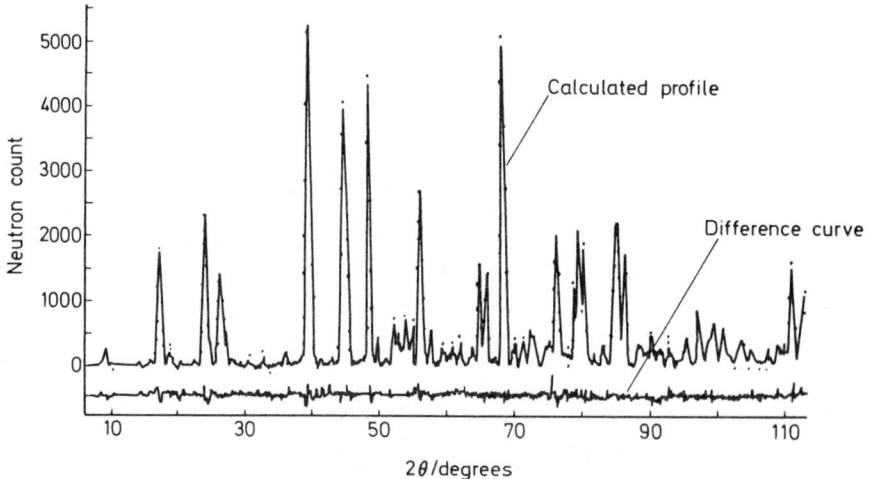

Figure 3.15 The powder diffraction pattern of $TiNb_2O_7$. The lines and points represent the calculated and observed values of the profiles, respectively, and the lower curve indicates the difference between them (From Van Dreele and Cheetham[13], by courtesy of Proc. Roy. Soc.)

intensity measurements respectively over the angular range from about 5° to 120° of 2θ. From a practical point of view it may be noted that the powdered material was contained in vanadium cans of 1.0 and 1.5 cm diameter and the diffraction pattern was traversed at a rate of 1 degree of 2θ in 50 minutes of time, corresponding to about 100 hours for covering the whole pattern. A general impression of the accuracy of the measurements and the resolution of the diffraction pattern may be gained from *Figure 3.15*. It will be appreciated from the data just given that when the method of profile refinement can be applied there is a great saving of time, in comparison with the time-consuming processes involved in the single-crystal technique of analysis.

References

1. MALISZEWSKI, E., SOSNOWSKI, J., BLINOWSKI, K., KOZUBOWSKI, J., PADLO, L. and SLEDZIEWSKA, D., in *Inelastic Scattering of Neutrons in Solids and Liquids*, Vol.II, 87, I.A.E.A., Vienna (1963)
2. BROCKHOUSE, B.N., in *Inelastic Scattering of Neutrons in Solids and Liquids*, 113, I.A.E.A., Vienna (1961)
3. IYENGAR, P.K., in *Thermal Neutron Scattering*, ed. P.A. Egelstaff, 97, Academic Press, London (1965)
4. KJEMS, J.K. and REYNOLDS, P.A., in *Neutron Inelastic Scattering*, 733, I.A.E.A., Vienna (1972)
5. WOODS, A.D.B., BROCKHOUSE, B.N., SAKAMOTO, M. and SINCLAIR, R.N., in *Inelastic Scattering of Neutrons in Solids and Liquids*, 487, I.A.E.A., Vienna (1961)
6. BURAS, B., *Nukleonika*, **8**, 259 (1963); BURAS, B., *AEC-ENEA Seminar*, Santa Fe, New Mexico (1967); BURAS, B. and LECIEJEWICZ, J., *Phys. Status Solidi*, **4**, 349 (1964); BURAS, B., LECIEJEWICZ, J., NITC, W., SOSNOWSKA, I., SOSNOWSKI, J. and SHAPIRO, F., *Nukleonika*, **9**, 523 (1964); BURAS, B., MIKKE, K., LEBECH, B. and LECIEJEWICZ, J., *Phys. Status Solidi*, **11**, 567 (1965)
7. SCHMATZ, W., SPRINGER, T., SCHELTEN, J. and IBEL, K., *J. Appl. Cryst.*, **7**, 96 (1974)
8. RIETVELD, H.M., *J. Appl. Cryst.*, **2**, 65 (1969)
9. LOOPSTRA, B.O. and RIETVELD, H.M., *Acta Cryst.*, **B25**, 787 (1969)
10. HEWAT, A., *J. Phys. C*, **6**, 2559 (1973); *Nature*, **246**, 90 (1973)
11. JACOBSON, A.J., COLLINS, B.M. and FENDER, B.E.F., *Acta Cryst.*, **B30**, 816 (1974)
12. JACOBSON, A.J., COLLINS, B.M. and FENDER, B.E.F., *Acta Cryst.*, **B30**, 1705 (1974)
13. VAN DREELE, R.B. and CHEETHAM, A.K., *Proc. Roy. Soc.*, **A338**, 311 (1974)
14. CAGLIOTI, C. and RICCI, F.P., *Nucl. Instrum. Methods*, **15**, 155 (1962)
15. AGERON, P., *Endeavour*, **113**, 67 (1972)
16. KIMURA, M., SUGAWARA, M., OYAMADA, M., YAMADA, Y., TOMIYOSHI,

S., SUZUKI, T., WATANABE, N. and TAKEDA, S., *Nucl. Instrum. Methods*, **71**, 102 (1969)
17. STIRLING, G.C., in *Chemical Applications of Thermal Neutron Scattering*, ed. B.T.M. Willis, Oxford University Press (1973)

4
STRUCTURAL STUDIES

Structural studies with single crystals were among the earliest applications of neutron diffraction and were developed as soon as the particular significance of secondary extinction on the intensities of diffraction had been realized through the work of Bacon and Lowde[1] and Peterson and Levy[2]. Applications based on the location of hydrogen atoms and the determination of the position of elements such as oxygen in heavy-element compounds were immediately evident. These have been pursued with increasing accuracy and over an ever-broadening range of materials, as the intensity of neutron beams has increased, making worthwhile, and indeed essential, the development of much more sophisticated experimental apparatus, capable of computer control for operation and adapted for rapid computer analysis of the data which are collected. In particular it is emphasized that the size of crystal needed for study has steadily fallen, from a linear dimension of about 6 mm initially to 0.6 mm at the present time, corresponding to a reduction of volume by a thousand times, in line with the increase of neutron flux which has been achieved in a quarter of a century. We shall discuss some typical analyses, mainly with single crystals, to indicate the scope of present methods. We emphasize that the basic techniques of data analysis are those of Fourier synthesis and least-squares analysis which were originally developed for X-ray crystallography. At the same time it is to be noted that corrections for secondary extinction, and for bond-shortening which occurs as a result of librations, are of substantially more importance for neutrons than for X-rays.

4.1 Sucrose: a typical study

As a good example of present methods of single-crystal analysis we will examine in some detail the study of sucrose, $C_{12}H_{22}O_{11}$, reported by Brown and Levy[3]. The intensity data for this work were collected in 1963[4] and consisted of about 5800 measurements. In order to make full corrections for secondary extinction, three separate crystals were examined, weighing respectively 96, 11 and 5 mg. For the largest crystal, which provided the majority of the data, 2813 independent reflections were examined. In the final refinement 425 parameters were determined, consisting of a scale factor and six anisotropic extinction factors for each crystal and three coordinates and six anisotropic thermal parameters for each of the 45 atoms in the asymmetric unit. *Figure 4.1* is a stereoscopic view of the molecule, indicating, in particular, the large anisotropic

40 Structural Studies

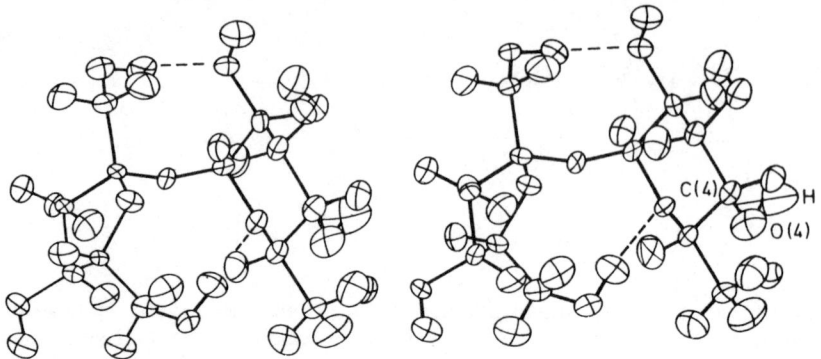

Figure 4.1 Stereoscopic view of the molecule of sucrose showing the 50% probability ellipsoids for the thermal motion. Note in particular the very large motion of the hydrogen atom attached to O_4 (From Brown and Levy[3], by courtesy of Acta Cryst.)

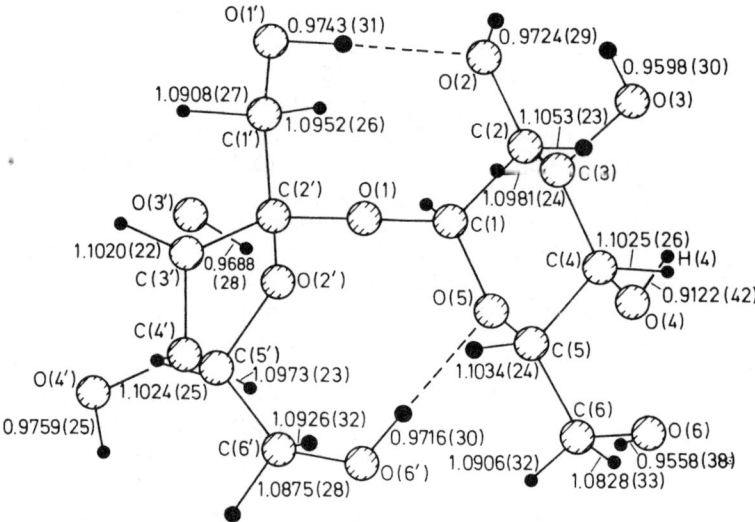

Figure 4.2 The structure of the molecule of sucrose, indicating the lengths of the C–H and O–H bonds. The small shaded circles, unlettered except for H(4), are the hydrogen atoms. There are eight different hydroxyl bonds in the molecule and, of these, five (which originate at O(2), O(3), O(6), O(3′) and O(4′)) take part in intermolecular hydrogen bonds. Two others, originating at O(1′) and O(6′), take part in intramolecular hydrogen bonds which are indicated by the broken lines. The remaining bond, between O(4) and H(4), is very short and does not take part in a hydrogen bond: the thermal motion of H(4) is exceptionally large, as can be seen in Figure 4.1 (After Brown and Levy[3], by courtesy of Acta Cryst.)

thermal motion found for the hydrogen atoms. The precision of the analysis, for which the conventional discrepancy factor R was 3.3%, is well illustrated by *Figure 4.2* on which the lengths and standard deviations of all the bonds which involve hydrogen atoms are marked. It will be seen that for C–H bonds the standard deviations vary between 0.0022 and 0.0033 Å and for O–H bonds they are between 0.0028 and 0.0042 Å. For C–C and C–O bonds the accuracy is significantly greater, with deviations between 0.0012 and 0.0019 Å. We emphasize that these bond lengths are the values calculated directly from the atomic coordinates without any correction for the shortening produced by anisotropic thermal motion, first discussed for neutrons by Busing and Levy[5] and summarized elsewhere[6].

It is evident from *Figure 4.1* that the large motion of the hydrogen atoms will lead to substantial apparent shortening of the bonds which involve hydrogen atoms. The magnitude of the corrections can only be assessed accurately when a full picture of the thermal motion within the molecule has been deduced. If, as a first approximation, it is assumed that the hydrogen atoms 'ride' on their carbon or oxygen atoms, then a correction can be computed. Thus, the average apparent length of the six C–H bonds within CH_2OH groups is 1.090 Å, whereas the average length of the other eight C–H bonds is 1.102 Å. When a 'riding' correction is made these two values become effectively indistinguishable at 1.122 and 1.121 Å, respectively. However, the riding model probably overcorrects, for the carbon atoms are not the centres of motion of the hydrogen atoms, and the true equilibrium value of the C–H separation is probably closer to 1.110 Å. The effect of making the bond-shortening correction to the O–H bonds is even more striking. There are eight different O–H bonds and in seven of these the hydrogen atom takes part in a hydrogen bond – of which two, shown by the broken lines in *Figure 4.2*, are *intra*molecular and the other five are *inter*molecular. These seven O–H bonds have uncorrected values which are quite closely similar, with an average value of 0.968 Å. On the other hand, the remaining O–H bond, between O(4) and H(4), is not involved in hydrogen bonding and is very much shorter, with an uncorrected value of 0.912 Å. However, the thermal motion of H(4) is exceptionally large, amounting to 0.43 Å, and when a riding correction is made the bond length is increased to 0.989 Å. This, albeit fortuitously, is found to be exactly the same as the corrected mean-length for the other seven O–H bonds.

4.2 Hydrogen bonds

One of the continuing topics of study in neutron crystallography has been the hydrogen bond. The first material whose structure was determined by single-crystal methods was KHF_2, closely followed by KH_2PO_4, which was of particular interest because of its ferroelectric properties, and then succeeded by a wide range of hydrated materials, particularly inorganic salts. The early workers were mainly concerned with the lengths of the components of the hydrogen bond, for example the O–H and O–O distances in an O–H⋯O bond, and many efforts were made to establish that there was a unique correlation between these lengths and it was believed that this correlation was concealed by inaccuracies in the early measurements. However, when higher reactor fluxes permitted more accurate determination of hydrogen positions it became evident that no single

42 *Structural Studies*

correlation existed and that the complete and detailed environment of the bonded atoms had to be taken into account. In many compounds the bonds are not linear and often their precise lengths are uncertain because of lack of detailed knowledge of the thermal motion of the atoms in the molecules. Nevertheless, even in circumstances where these two features do not seem to set limiting conditions it cannot be concluded at the present time that the geometry of hydrogen bonds is satisfactorily understood.

In order to indicate the accuracy and scope of present studies of hydrogen bonds we shall describe the work on very short bonds. We recall that KHF_2, to which we referred above, contains the shortest hydrogen bond known, measuring 2.26 Å and deduced to be a centred bond by Peterson and Levy[7]. A similar short bond was later studied in $NaHF_2$ by McGaw and Ibers[8]. Among the first-known short bonds in organic compounds is that now accepted as 2.454 Å in potassium hydrogen bisphenylacetate, a compound which has been studied with both X-rays and neutrons, culminating in three-dimensional analyses by Manojlovic and Speakman[9] with X-rays and Bacon, Walker and Speakman[10] using neutrons: again a centred bond is believed to exist. As a result of the work on these and other substances there has been a tendency to identify the short bonds with symmetric centred bonds, but it is now known that this is incorrect. Thus Williams, Peterson and Levy[11] have shown that in sulphosalicylic acid trihydrate there is a very unsymmetrical short bond of length 2.436 Å, in which the two sections O—H and H⋯O have lengths of 1.095 and 1.341 Å. As *Figure 4.3* shows, although the hydrogen atom lies between two water molecules (to form a $[H_5O_2]^+$ ion), the external environments of these two water molecules are quite different. Even more striking is

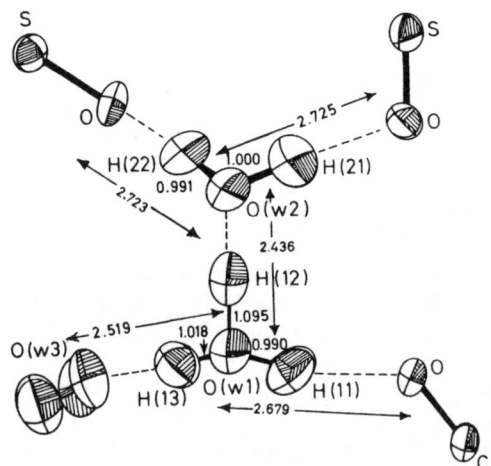

Figure 4.3 The hydrogen bonds in sulphosalicylic acid trihydrate. The environments of the water molecules w1 and w2 are quite different, with the result that the hydrogen bond between them, from O(w1) → H(12) → O(w2) is, although short, very unsymmetrical, with the two sections measuring 1.095 and 1.341 Å (From the results of Williams, Peterson and Levy[11])

the very short F—H—F bond of length 2.26 Å which is present in p-toluidinium bifluoride (Williams and Schneemeyer[12]). Here the two sections measure 1.025 and 1.235 Å and, as *Figure 4.4* shows, the environments at the two ends of the bond F(1)—F(2) are quite different. F(1) is much more rigidly bound, with hydrogen bonds of 1.607 and 1.675 Å, whereas F(2) takes part in only one other hydrogen bond, of length 1.777 Å. At the same time, as would be expected, the thermal motion of F(2) is much larger than that of F(1). This is emphasized in the ellipsoids of *Figure 4.5*.

It will be noted further in *Figure 4.3* that the three O—H bonds within the oxonium ion are of significantly different lengths because of the different environments which are found in these three directions. In particular the two O—H bonds from O(w1) to H(11) and H(13) measure 0.990 and 1.018 Å, respectively, being joined to O(7) and a further water molecule O(w3) in the two cases. By contrast, in the structure of p-toluene-sulphonic acid monohydrate (Lundgren and Williams[13]) (*Figure 4.6*) the environment of the central $[H_3O]^+$ ion is the same in each of three directions, there being an oxygen of a sulphonate group in each direction. In

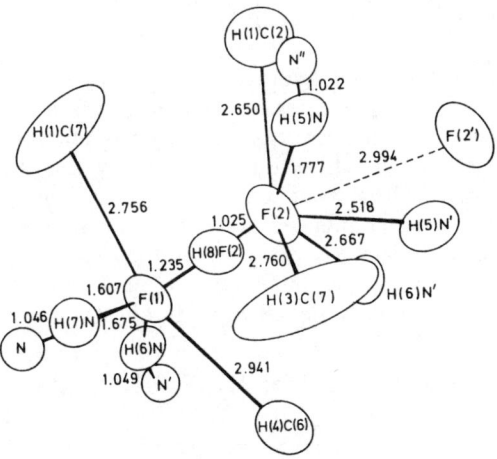

Figure 4.4 The very short unsymmetrical hydrogen bond between two fluorine atoms F(1) and F(2) in p-toluidinium bifluoride. The environments of F(1) and F(2) are quite different and the two sections of the bond measure 1.235 and 1.025 Å, respectively (From Williams and Schneemeyer[12], by courtesy of J. Amer. Chem. Soc.)

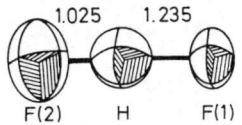

Figure 4.5 Ellipsoids of thermal motion for the fluorine atoms F(1) and F(2) in p-toluidinium bifluoride which were shown in Figure 4.4. F(1) is much more rigidly bound in the structure and its motion is correspondingly less (From Williams and Schneemeyer[12], by courtesy of J. Amer. Chem. Soc.)

44 Structural Studies

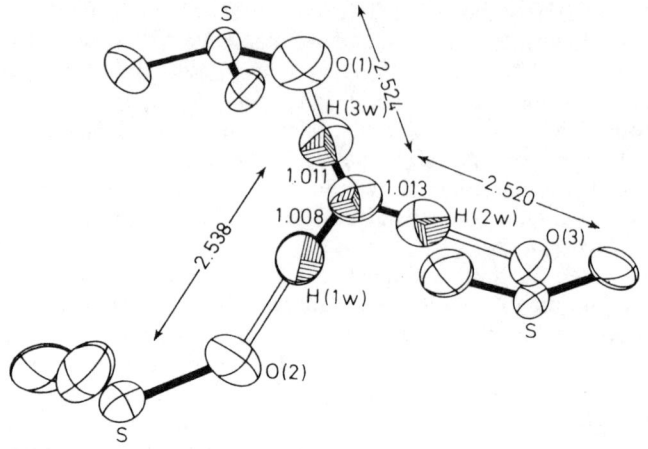

Figure 4.6 Bond lengths for the $(H_3O)^+$ ion in p-toluenesulphonic acid monohydrate $[H_3O]^+[CH_3C_6H_4SO_3]^-$. The ion is similarly bonded in three directions to oxygen atoms of sulphonate groups and the three hydrogen bonds are indistinguishable in dimensions within experimental error (After Lundgren and Williams[13], by courtesy of J. Chem. Phys.)

turn it is found that the three hydrogen bonds are indistinguishable, within experimental error, and the three O–H bonds measure 1.009, 1.013 and 1.011 Å, respectively. Both the ordinary p-toluenesulphonic acid monohydrate and the material in which the oxonium ion has been deuterated (Finholt and Williams[14]) have been examined and a comparison of the two studies provides a good indication of the accuracy of the conclusions from neutron diffraction analysis for this kind of molecule. The notable parameters in the two materials are compared in *Table 4.1*. It will be seen that deuteration produces no change in the average value of the O–H

Table 4.1 Comparison of results for H_3O^+ $CH_3C_6H_4SO_3^-$ and D_3O^+ $CH_3C_6H_4SO_3^-$

	Hydrogen			Deuterium		
Reliability index of analysis/%	6.7			5.3		
Distance from O to base of $[H_3O]^+$ pyramid/Å	0.322 (4)			0.334 (3)		
RMS amplitudes of vibration for methyl group/Å	0.210 0.317 0.448	0.212 0.325 0.458	0.212 (9) 0.299 (10) 0.487 (13)	0.209 0.311 0.455	0.207 0.335 0.433	0.199 (7) 0.309 (7) 0.504 (9)
O–O bond lengths/Å		2.529 (8) 2.545 (8) 2.522 (8)			2.537 (4) 2.549 (4) 2.537 (4)	
mean value/Å		2.532			2.541	
O–H bond lengths/Å		1.013 (8) 1.011 (8) 1.014 (8)			1.008 (3) 1.011 (3) 1.020 (3)	
mean value/Å		1.013			1.013	

distance in the oxonium ion. There is for each of the three bonds an increase in the overall O—O distance between the oxonium and sulphonate ions, but the changes are small enough to be of doubtful significance. There is also an increase in the height of the rather flat pyramid which constitutes the oxonium ion but, again, this is barely significant in magnitude. The closely similar values found for the amplitudes of vibration of the hydrogen atoms in the methyl group in the two materials suggest that the estimated errors in these amplitudes are by no means over-optimistic. The methyl groups undergo a torsional vibration of large amplitude about the C—C bonds which join them to the benzene ring.

These two sets of measurements were made with crystals of about 17 and 13 mm^3 with a medium-flux reactor and provide a good example of the chemical conclusions which may be reached under these circumstances. They also support the general conclusion that the isotope effect, so far as it determines bond lengths in hydrogen-bonded systems, is very small except when metal–oxygen bonds are involved. Thus in the case of $HCrO_2$ (Hamilton and Ibers[15]) and $HCoO_2$ (Delaplane et al.[16]) deuteration leads to an increase in the O—O bond lengths of 0.06 and 0.07 Å, respectively, whereas in the work which we have just described for the p-toluenesulphonic acids, which is quite typical, the changes were only 0.008, 0.004 and 0.015 for the three distinct O—O bonds.

The general development of neutron diffraction techniques, and in particular their application to the study of hydrogen bonds, is well illustrated by the studies of ice, which was first examined in 1949 as a powder by Wollan, Davidson and Shull[58] and then as a single crystal by Peterson and Levy[59]. The latter, using deuterated material, established that there was a disordered distribution of the protons among the pairs of possible positions between the oxygen atoms. Subsequently much further work has been done, supplementing X-ray work, to elucidate the structures of the polymorphs of ice which are formed at high pressures. Results have been obtained[60,61], mainly for deuterated ice, for both powders and single crystals in which the structures have been frozen-in at low temperatures to permit easier examination at normal pressures. A summary of this work was given[62] by Speakman in 1973.

As our final example of studies of hydrogen bonds we mention the work of Brunton and Johnson[63] on yttrium oxalate trihydrate. This has confirmed the suggestion, from earlier X-ray work, that an $H_5O_2^+$ ion is present, with a proton nearly centred between two water molecules. By making measurements with partially deuterated materials of various H/D isotopic ratios the neutron data established that there was a predominance of hydrogen, rather than deuterium, in this centred site and the predominance increased at reduced temperatures.

4.3 Molecular overcrowding

Another topic on which the accurate location of hydrogen atoms provides convincing information is the study of changes in molecular shape to offset the effects of overcrowding. A typical example is phenanthrene,

46 Structural Studies

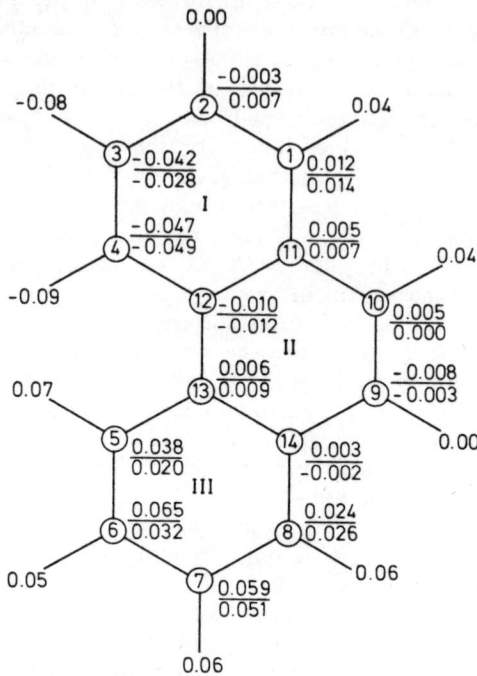

Figure 4.7 Overcrowding of the hydrogen atoms linked to carbon atoms C(4) and C(5) in the molecule of phenanthrene, $C_{14}H_{10}$. Distances are shown above or below the average plane through the ring II of carbon atoms. For the carbon atoms values are given for both neutrons (upper) and X-rays (lower): for the hydrogen atoms only neutron data are given. The upper ring I is twisted about the line C(2)C(12) to take C(3), C(4) and H(3), H(4) below the plane of the paper, whereas ring III is twisted in the contrary direction, bringing C(5), C(6) and H(5), H(6) above the paper (From Kay, Okaya and Cox[17], by courtesy of Acta Cryst.)

$C_{14}H_{10}$, which has been studied with both X-rays and neutrons by Kay, Okaya and Cox[17] and for which the molecule is shown in *Figure 4.7*. At first thought it would be expected that the molecule would be planar; if this were the case, with normal bond lengths and bond angles, then the hydrogen atoms which are attached to C(4) and C(5) would be only about 1.8 Å apart. This distance is substantially shorter than the van der Waals' distance for two hydrogen atoms which has been variously estimated between 2.0 and 2.4 Å and, accordingly, the molecule distorts from its ideal planar shape in order to reduce the extent of the overcrowding. The neutron data show that twisting of the molecule increases the H–H separation to 2.04 Å. The twisting is evident from *Figure 4.7* in which are marked the distances of the various atoms from the mean plane through the carbon atoms of ring II. For the carbon atoms out-of-plane distances are shown from both the X-ray and the neutron study, but for the hydrogen atoms only the distances from the neutron study are shown. It is evident that the top ring (I) is twisted about the C(2)–C(12) axis, taking atoms C(3), C(4), H(3) and H(4) below the plane of the paper. On the other hand, C(5), C(6), H(5) and H(6) are brought above the plane of the paper by a contrary twist of ring III about the C(13)–C(7) axis and a further twist about C(13)–C(14). Although the distortion is very evident, the precision of its description in detail is limited by the correction for effective bond-shortening brought about by anisotropic thermal motion. The thermal parameters of the carbon atoms are consistent with the assumption that each carbon ring moves as a rigid body, but the hydrogen atoms do not conform to this and show substantial internal motion for which the correction for bond-shortening is difficult to assess precisely. Measurements at low temperature, where the thermal motion is reduced, would be likely to clarify the details of the distortion of the molecule.

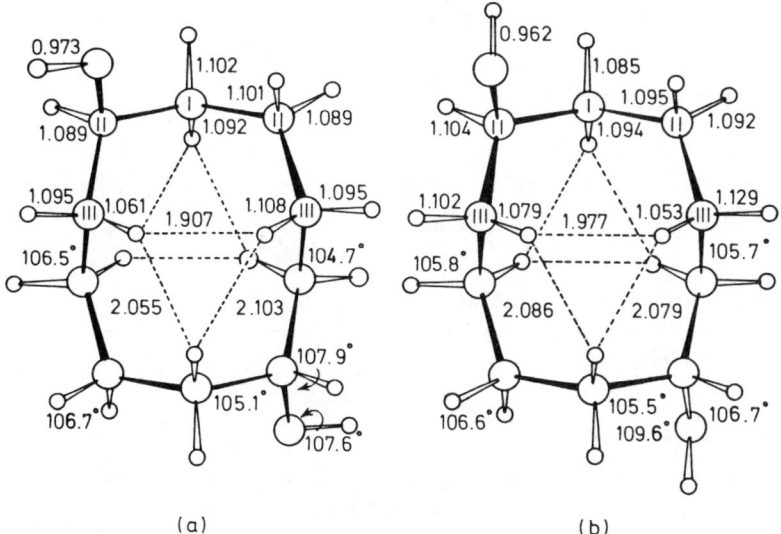

Figure 4.8 A plan showing the two crystallographically different molecules in the structure of trans-cyclodecane-1,6-diol, $C_{10}H_{18}(OH)_2$. The two molecules (a) and (b) differ in the disposition of the hydrogen atoms of the hydroxyl groups relative to the C_{II}–O bonds in the top-left and bottom-right corners of the molecules. In each case there are unusually short distances across the ring between the pairs of hydrogen atoms attached to carbon atoms of type III. These distances, of 1.907 and 1.977 Å respectively, are larger than would be achieved in a conventional carbon framework and have been achieved by distortions within the molecule of up to 5° (From Ermer and Dunitz[18], by courtesy of Chem. Commun.)

A somewhat similar situation has been studied by Ermer and Dunitz[18] in *trans*-cyclodecane-1,6-diol, $C_{10}H_{18}(OH)_2$. In this substance there are two crystallographically different molecules, each of which is centrosymmetrical but which show different conformations for the hydroxyl groups. In each case a pair of very short transannular H–H distances is observed, amounting to 1.907 ± 0.008 Å and 1.977 ± 0.009 Å in the two cases, as illustrated in *Figure 4.8*. These distances are substantially shorter than the van der Waals' distance of 2.4 Å, but significantly longer than the calculated values of 1.80 and 1.92 Å respectively which would be expected for a conventional carbon framework and methylene groups, for

48 Structural Studies

which the C–H bonds would be 1.10 Å and the HCH angle would be 106°. The angles within the molecule are found to be distorted by up to 5° in an attempt to reduce the overcrowding of the hydrogen atoms within the ring. It is also noteworthy that both the thermal motion and its anisotropy are smaller for these hydrogen atoms than for the extra-annular hydrogen atoms.

Finally we record the case of $C_{18}H_{18}$, [18] annulene, often discussed by the late C.A. Coulson, in which the six hydrogen atoms which are inwardly directed from the carbon framework would be permitted interatomic distances of only 1.7 Å if the molecule remained planar. Buckling of the molecule has been demonstrated with X-rays (Bregman et al.[19], Hirshfeld and Rabinovich[20]), but a neutron study, to locate the hydrogen atoms precisely, has not yet been carried out.

4.4 Heavy-element compounds

An excellent example of the value of neutron diffraction for studying the structures of compounds which include heavy elements is provided by mercury chromate hemihydrate, $HgCrO_4 \cdot \tfrac{1}{2}H_2O$, which has been studied with neutrons by Aurivillius and Stålhandske[21], following an earlier X-ray study

Table 4.2 Scattering lengths (in 10^{-12} cm) for $HgCrO_4 \cdot \tfrac{1}{2}H_2O$

Atom	Hg	Cr	O	H
X-ray	22.5	6.8	2.2	0.3
Neutron	1.26	0.36	0.58	−0.37

Table 4.3 Comparison of accuracies of position parameters for neutrons (upper) and X-rays (lower) for $HgCrO_4 \cdot \tfrac{1}{2}H_2O$

Atom	\multicolumn{2}{c}{Fractional coordinates}				
	x		y		z	
Hg	−0.00110	(5)	0.02619	(9)	0.12474	(4)
	−0.00126	(10)	0.02676	(12)	0.12468	(8)
Cr	0.1995	(1)	0.4391	(3)	0.0983	(1)
	0.1999	(3)	0.4385	(4)	0.0989	(2)
O(1)	0.3671	(1)	0.3913	(2)	0.1657	(1)
	0.3687	(17)	0.3925	(25)	0.1646	(14)
O(2)	0.1629	(1)	0.6174	(2)	0.1683	(1)
	0.1610	(20)	0.6187	(31)	0.1660	(16)
O(3)	0.1213	(1)	0.1514	(2)	0.0732	(1)
	0.1195	(17)	0.1562	(28)	0.0728	(14)
O(4)	0.1514	(1)	0.5733	(2)	−0.0141	(1)
	0.1494	(20)	0.5757	(35)	−0.0161	(16)
O(w)	0		0.3967	(3)	¼	
			0.3989	(44)		
H	0.0725	(3)	0.5076	(6)	0.2878	(2)

by Aurivillius[22]. *Table 4.2* indicates how the relative values of the scattering lengths of the four elements in this compound are quite different for the two radiations. These differences are reflected in *Table 4.3*, which gives the fractional atomic coordinates deduced in the two cases, together with their estimated standard deviations. It is noteworthy that the hydrogen position could be determined only with neutrons and that the accuracy of the oxygen positions is ten to twenty times better with neutrons than for X-rays. The structure consists of endless chains, of composition $(HgCrO_4)_n$, but in the X-ray study it was only possible to speculate, and incorrectly, how they were held together by the water molecules. In the neutron study the water molecules are clearly defined: they lie in holes between the chains, which they bind together by weak hydrogen bonds to produce parallel slabs of material with a thickness of $c/2$. This is made evident in *Figure 4.9*, which shows how the chains are joined by the hydrogen bonds. The bonds are weak and long, with a distance of 2.96 Å between the water-oxygen and the oxygen of the chromate group and the bonds are very bent, with an O–H···O angle of only 146°. It is interesting to note that the X-ray study measured 1702 reflections from a spherical crystal of volume 0.07 mm^3, whereas the neutron study, utilizing the medium flux reactor at Studsvik, Sweden, measured 1084 reflections from a crystal of 5.5 mm^3.

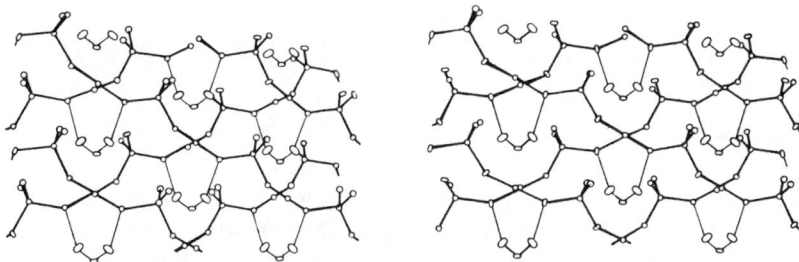

Figure 4.9 A stereoscopic view of the structure of $HgCrO_4 \cdot \tfrac{1}{2}H_2O$ indicating how the water molecules occupy holes between endless chains of composition $(HgCrO_4)_n$ and bind these together by weak hydrogen bonds (From Aurivillius and Stålhandske[21], by courtesy of Z. Krist.)

The structures of compounds of uranium and other actinide elements are very difficult to determine with X-rays because of the immense scattering power of these elements compared with, say, oxygen or nitrogen. Thus the scattering amplitudes of uranium and oxygen for X-rays are approximately 26×10^{-12} and 2×10^{-12} cm, respectively. For neutrons, on the other hand, the scattering length of uranium is only about 50% greater than that of oxygen, the values being 0.85×10^{-12} and 0.58×10^{-12} cm, respectively. The advantage of neutrons can be seen in the analysis of the complicated structure of U_4O_9 by Masaki and Doi[23], using single crystals having linear dimensions up to 5 mm. The structure of U_4O_9 is based on that of UO_2, with a unit cell which is cubic with an edge of 21.76 Å. This length is four times that of the cell of the parent substance. Accordingly 64 U_4O_9 'molecules' have to be accommodated within the space which is normally occupied by 64 U_4O_8 'molecules', so that room has to be found for 64 extra oxygen atoms. Experimentally it is found that the fundamental reflections associated with the basic fluorite structure are accompanied by weak superlattice reflections from the enlarged unit cell, suggesting long-range order in the arrangement of the additional oxygen atoms. The superlattice reflections which are observed, and their intensities, are indicated by the black dots in *Figure 4.10*, which is a section of the reciprocal lattice in

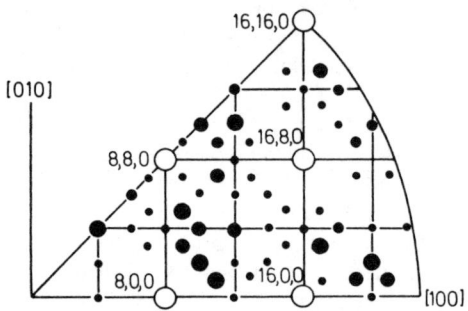

Figure 4.10 A section of the reciprocal lattice of U_4O_9 by the xy plane, with the magnitudes of the black spots to indicate the approximate magnitude of the superlattice reflections. The fundamental reflections, indicated by the open circles, have intensities which are of the order of 1000 times as large (From Masaki and Doi[23], by courtesy of Acta Cryst.)

the 001 plane. The intensities of most of the superlattice reflections are only about one-thousandth of those of the fundamental reflections, whose positions are marked by the open circles in the figure. By a detailed analysis of the intensities of the superlattice reflections and a construction of Fourier difference maps, from the superlattice reflections alone, it has been possible to find the details of the structure. It is found that 40 of the extra oxygen atoms are in interstitial positions at the centres of the cubelets of oxygen atoms, of edge 2.73 Å, which characterize the underlying UO_2 structure. The remaining 24 of the additional oxygen atoms are displaced from cubelet centres by 0.52 Å, in the directions of face diagonals. At the same time, in order to accommodate these movements in the oxygen framework, 16 of the 256 uranium atoms in the large unit cell are displaced from the positions which they normally occupy. The displacement is along the direction of body-diagonals and amounts to 0.45 Å. These changes are indicated in *Figure 4.11*, which shows a projection of one layer of the enlarged unit cell, consisting therefore of 4 × 4 × 1 of the original UO_2 cells, in which only the additional oxygen atoms and the displaced uranium atoms are shown. The arrows, and their accompanying zone-axis symbols, indicate the directions of displacement of the uranium atoms and of those oxygen atoms which belong to the displaced group of 24. The figures alongside the atomic circles indicate the c coordinates in terms of the

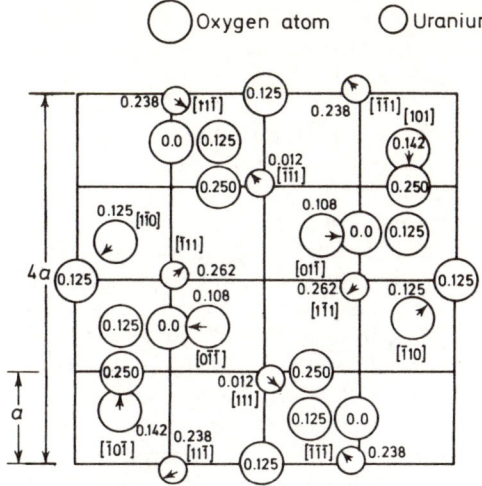

Figure 4.11 The arrangement of the additional oxygen atoms (large circles) and the displacements of the uranium atoms (small circles) in one layer of the enlarged unit cell postulated for U_4O_9, projected on the 001 plane. The figures indicate the c coordinates of the atoms in terms of the enlarged unit cell. The additional oxygen atoms are either at the centres of cubelets or displaced by 0.52 Å along a face diagonal as indicated by the arrows. The uranium atoms shown are displaced by 0.45 Å along the body diagonal specified by the zone-axis symbol (From Masaki and Doi[23], by courtesy of Acta Cryst.)

enlarged cell. In the absence of any displacements these coordinates would be 0, 0.125 or 0.250 for the oxygen atoms and 0 or 0.125 for the uranium atoms.

Among other applications to heavy-element compounds are the studies of the scheelites $CaMoO_4$, $SrMoO_4$, $SrWO_4$ and $BaWO_4$. For neutrons the scattering amplitudes of the atoms in these compounds are all quite similar, whereas with X-rays the oxygen atoms are very much outweighted by the metal atoms. In the original X-ray investigations of Sillen and Nylander[24] the precision of the determination of the parameters x, y, z defining the positions of the oxygen atoms is indicated by the results for $CaWO_4$, which gave

$x = 0.25 \pm 0.02$, $y = 0.15 \pm 0.02$, $z = 0.075 \pm 0.015$

Later X-ray work by Zalkin and Templeton[25] measuring 828 reflections gave

$x = 0.2415 \pm 0.0014$, $y = 0.1504 \pm 0.0013$, $z = 0.0861 \pm 0.0006$

A neutron study by Kay, Frazer and Almodovar[26] measuring only 57 reflections in the $h0l$ and hhl zones at a quite low-flux reactor gave

$x = 0.2413 \pm 0.0005$, $y = 0.1511 \pm 0.0006$, $z = 0.0861 \pm 0.0001$

Further improvements in precision could be gained from a standard three-dimensional analysis at a high-flux reactor. It is interesting to note that $CaWO_4$ is a very favourable special case. Calcium and tungsten have neutron scattering amplitudes which are almost identical and their atomic positions have x,y coordinates which are identical and z coordinates which differ by ½. This means that for hkl reflections for which l is odd there is practically no contribution at all from the metal atoms. Consequently the parameters of the oxygen atoms can be found very accurately.

We have already described in Chapter 3, in our general description of the method of powder-profile refinement, the work of Loopstra and Rietveld[27] using powdered samples of $CaUO_4$, $SrUO_4$, $BaUO_4$, Ca_2UO_5, Sr_2UO_5 and Ca_3UO_6. As later examples of the use of profile refinement for the study of uranium compounds we mention the work of Taylor and Wilson[28] on UF_6 and of Levy, Taylor and Wilson[29] on UI_3. In the former case measurements were made at temperatures of 193 and 293 K. The precision of the analysis is significantly less at the higher temperature, no doubt partially due to the assumption of a single overall isotropic temperature-factor, but it was concluded that the UF_6 octahedra become substantially more regular on cooling.

A quite different field of uranium chemistry has been explored by Leciejewicz and his co-workers[30,31], who have studied such materials as UAs, USe, UAsSe, USb_2 and various solid solutions. A number of these substances have possible applications in nuclear technology and have been studied mainly from the point of view of their magnetic properties. As typical examples we may consider UAs and its solid solutions which incorporate sulphur or selenium: these materials all have the rock-salt type of crystallographic structure. UAs itself was shown by Leciejewicz, Murasik and Troc[30] to be antiferromagnetic with a Néel temperature of 128 K and a magnetic structure at lower temperatures, as indicated in *Figure 4.12(a)* and in which the uranium atoms possess a magnetic moment of 1.93 μ_B. Below 66 K there is a transition to the new structure shown in *Figure 4.12(b)*. The unit cell is now doubled in the direction of the c axis and is tetragonal: the magnetic moments of the uranium atoms are now 2.20 μ_B. For USb no such magnetic transition takes place and a structure like that of *Figure 4.12(a)* persists down to 4 K. It has been found by Leciejewicz *et al.*[31] that the low-temperature phase of UAs can be greatly stabilized by replacement of some of the arsenic by sulphur, though it is to be noted that US itself is ferromagnetic. Thus at the composition $UAs_{0.85}S_{0.15}$ the structure of *Figure 4.12(b)* persists up to a temperature of 115 K. At somewhat lower temperatures the structure can be observed up to a sulphur content indicated by $UAs_{0.65}S_{0.35}$, but here it is also accompanied by a further antiferromagnetic structure which has a unit cell which is five times as large, in the c-direction, as the original rock-salt cell and which acts as a transient phase for the transfer to the ferromagnetic phase which is solely present for sulphur compositions greater than that of $UAs_{0.60}S_{0.40}$. In this ferromagnetic phase the magnetic moment is 1.70 μ_B. In the intervening five-unit structure the z component of the magnetic moment varies sinusoidally along the c-axis with a modulation period of 2.5 a. For more details of

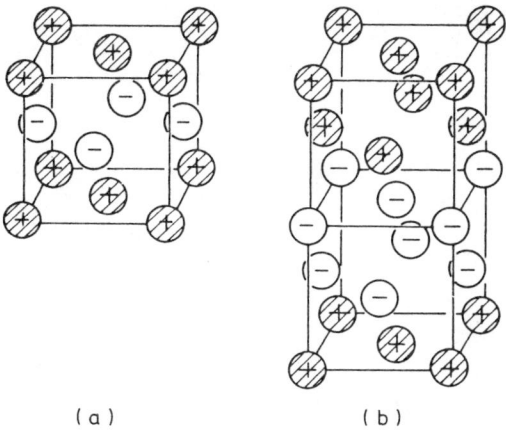

Figure 4.12 Diagram (a) indicates the uranium atoms and their magnetic moment directions in the rock-salt structure of UAs between 66 and 128 K. Below 66 K there is a transition to the structure shown in (b) with a unit cell which is doubled in the direction of the c axis. For USb, structure (a) persists down to 4 K

the way in which such magnetic structures are revealed in the neutron diffraction pattern, and of the variety of structures which have been found in various materials, the reader is referred to more general accounts of magnetic crystallography (for example, Bacon[6]) and to the original papers of Leciejewicz and his co-workers[30,31].

4.5 Some simple molecules

There is a number of simple compounds such as HCl and H_2S whose structures have not been adequately explored with X-rays because the hydrogen atoms, whose positions and motions are so important to a chemical understanding, cannot be located sufficiently accurately. Many of these have now been examined with neutrons, particularly by Sándor and his co-workers. Most of these substances are liquids or gases at room temperature so that their solid structures are necessarily examined at reduced temperature: it is generally found that the experimental difficulties in doing this are less severe with neutrons than for X-rays. The accuracy of the conclusions which can be reached with neutrons often depends on the resolution which can be achieved in powder-diffraction patterns. Higher-flux reactors have led to modification of conclusions reached in the earliest experiments with some of these materials.

As an example of what has been learnt concerning the structures which these simple molecules form, we will examine first HCl and its deuterated form DCl. The latter is much more attractive for examination with neutrons because of the absence of troublesome incoherent scattering. These materials undergo structural transitions at 98.4 and 105 K, respectively, and below these temperatures the crystals are ferroelectric. Sándor and Farrow[32,33] deduced from the neutron diffraction patterns of

54 Structural Studies

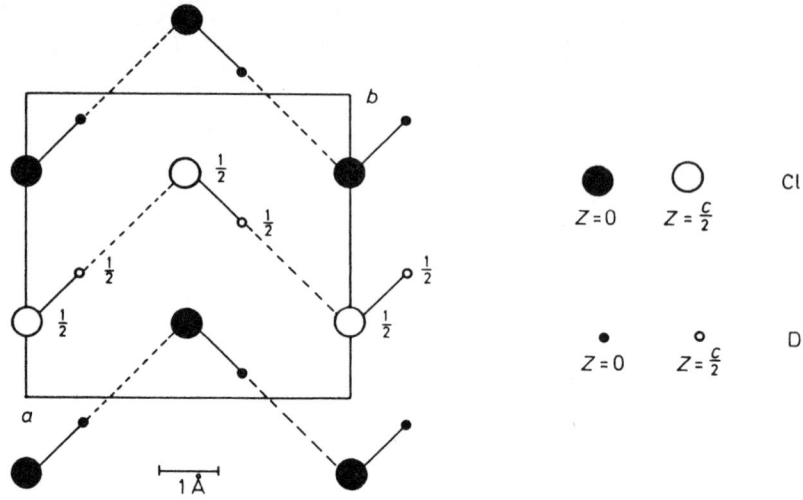

Figure 4.13 Arrangement of molecules in the ordered low-temperature orthorhombic phase of DCl (From Sandor and Farrow[34], by courtesy of Discuss. Faraday Soc.)

powdered samples that DCl had an orthorhombic structure of space group $Bb2_1m$ below the transition temperature and a cubic structure, space group $Fm3m$, at higher temperatures. *Figure 4.13* shows the structure of the low-temperature phase, which is characterized by planar zig-zag chains of hydrogen bonds, with the chains lying in the 001 planes. When the transition to the face-centred cubic structure takes place there is little

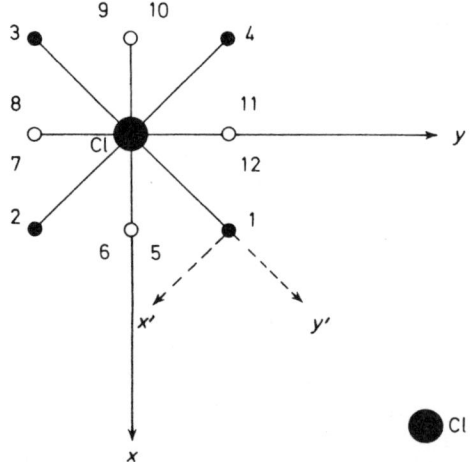

Figure 4.14 Arrangement of molecules in the disordered cubic phase of DCl. The twelve small circles, numbered 1-12, indicate the twelve possible positions for the deuterium atom around the chlorine atom indicated by the large circle. The filled small circles are in the plane of the drawing and the open circles are sites above and below the paper. The arrows labelled x' and y' indicate the first two principal axes of the thermal vibration tensor of the deuterium atom at site 1 (From **Sandor** and Farrow[34], by courtesy of Discuss. Faraday Soc.)

change in the position of the chlorine atoms but some randomness occurs in the orientation of the D–Cl molecules and a 'twelve-fold disordered' model is proposed. The deuterium atom of each molecule has twelve equally probable positions, situated along the lines which join any chlorine atom to its twelve nearest neighbours. The twelve positions are enumerated in *Figure 4.14*, where 1, 2, 3 and 4 are in the plane of the paper and (5,6), (7,8), (9,10) and (11,12) represent pairs of atoms which are respectively above and below the paper. In a later paper Sándor and Farrow[34] have studied the molecular motion in the two phases of DCl and we shall describe their conclusions in some detail because they illustrate how the knowledge of position and motion of H or D atoms can be extremely definitive.

The atomic motions of D and Cl will be compounded from translational and angular motion of the DCl molecule as a rigid body and the internal stretching of the molecule, which gives a relative displacement of the D and Cl atoms. The stretching vibration may be regarded at low temperature as the zero-point vibration of a harmonic oscillator and the mean-square displacement $\langle u_s^2 \rangle$ is related to the frequency ν_s observed in the Raman spectrum by

$$\langle u_s^2 \rangle = h/8\pi^2 \mu \nu_s \qquad (4.1)$$

where h is Planck's constant and μ is the reduced mass of the molecule, equal to $m_D m_{Cl}/(m_D + m_{Cl})$. The displacements for D and Cl are then given by the two equations

$$\langle u_s^2(Cl) \rangle = [m_D/(m_{Cl} + m_D)]^2 \langle u_s^2 \rangle$$

and $\quad \langle u_s^2(D) \rangle = [m_{Cl}/(m_{Cl} + m_D)]^2 \langle u_s^2 \rangle \qquad (4.2)$

with the results listed in *Table 4.4* for measurements at 10, 77 and 133 K, respectively. At the same time the anisotropic temperature factors deduced from the neutron diffraction intensities will depend on both the internal and external vibrations of the molecule. These are indicated in *Table 4.5*, from which we deduce that the principal axes of the chlorine motion coincide with the three edges of the unit cell and that the third axis of the deuterium motion coincides with the c-axis. *Table 4.4* then demonstrates that the stretching vibration contributes insignificantly to the chlorine motion but is significant for the deuterium atoms. Consequently the values of U_{11}, U_{22} and U_{33} for chlorine in *Table 4.5* are identical with the mean-square *translational* displacements $\langle t_a^2 \rangle$, $\langle t_b^2 \rangle$ and $\langle t_c^2 \rangle$ for the molecule as a whole. These displacements are therefore known. In turn, the values of U_{11}, U_{22} and U_{33} for deuterium can, after subtraction of the planar contributions from the D–Cl stretching, be equated with the sum of the translational and rotational motion of the molecule according to the three equations

Table 4.4 Mean square stretching displacements for DCl

T/K	Phase	ν_s/cm^{-1}	$\langle u_s^2 \rangle$	$\langle u_s^2 (Cl) \rangle$ /10^{-2} Å^2	$\langle u_s^2 (D) \rangle$
10	orthorhombic	1977	0.446	0.001	0.400
77	orthorhombic	1982	0.446	0.001	0.400
133	cubic	2012	0.440	0.001	0.394

Table 4.5 Thermal parameters of DCl from neutron diffraction (10^{-2} Å^2)

T/K		4.2	77.4	92.4	115.5
			orthorhombic		cubic
Cl	U_{11}	0.4	3.5	3.4	4.8
	U_{22}	3.0	2.6	2.1	4.8
	U_{33}	0.5	2.0	2.6	4.8
	U_{12}	0	0	0	0
	U_{13}	0	0	0	0
	U_{23}	0	0	0	0
D	U_{11}	1.3	5.0	6.2	8.1
	U_{22}	3.9	4.0	4.7	8.1
	U_{33}	2.4	4.0	4.5	20.2
	U_{12}	1.5	−1.4	1.1	−4.5
	U_{13}	0	0	0	0
	U_{23}	0	0	0	0

$$(U_{11})_D = \langle t_a^2 \rangle + L^2 \langle \phi_{in}^2 \rangle \cos^2 a = (U_{11})_{Cl} + L^2 \langle \phi_{in}^2 \rangle \cos^2 a$$

$$(U_{22})_D = \langle t_b^2 \rangle + L^2 \langle \phi_{in}^2 \rangle \sin^2 a = (U_{22})_{Cl} + L^2 \langle \phi_{in}^2 \rangle \sin^2 a$$

$$(U_{33})_D = \langle t_c^2 \rangle + L^2 \langle \phi_{out}^2 \rangle = (U_{33})_{Cl} + L^2 \langle \phi_{out}^2 \rangle \quad (4.3)$$

Here L is the length of the Cl–D bond, a is the inclination of this bond to the x axis and $\langle \phi_{in}^2 \rangle$ and $\langle \phi_{out}^2 \rangle$ are the mean square displacements of the in-plane and out-of-plane angular vibrations of the molecule. From these equations the values of $\langle \phi_{in}^2 \rangle$ and $\langle \phi_{out}^2 \rangle$ were determined at each temperature. The results for both the translation and rotation of the molecules are indicated in *Figure 4.15*, which shows two interesting features. First the zero-point vibration of the molecule has a considerable translational component along the *b*-axis, which is the direction of the chains shown in *Figure 4.13*, and considerable angular components $\langle \phi_{in}^2 \rangle$ and $\langle \phi_{out}^2 \rangle$. Secondly we see that the in-plane angular vibration $\langle \phi_{in}^2 \rangle$ increases very rapidly as the transition temperature is approached. This suggests that the orientational disorder of the molecules, which is present in the cubic phase as we mentioned in relation to *Figure 4.14*, may start below the transition temperature as the molecules reorient themselves in the plane of the chains.

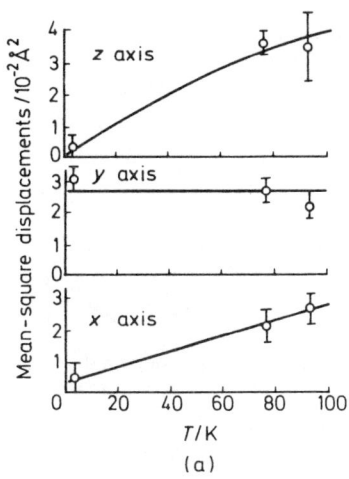 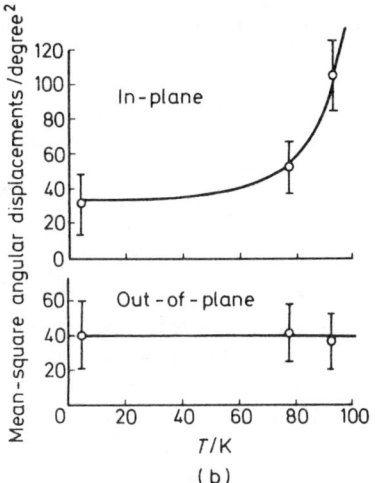

Figure 4.15 The variation with temperature of (a) the mean-square translational displacements and (b) the mean-square angular displacements, of the molecules in the ordered orthorhombic phase of DCl. For the translations the three curves show respectively motion along the x, y and z axes. For the rotation the curves are for the in-plane and out-of-plane motion (From Sandor and Farrow[34], by courtesy of Discuss. Faraday Soc.)

A similar assessment of the thermal motion can be made for the disordered cubic phase of DCl, bearing in mind that there are twelve possible equivalent orientations of the molecule. In terms of the particular position numbered 1 in *Figure 4.14* it is found that the displacement out of the plane (i.e. along z) is much larger than the displacement in the xy plane (caused by rotation about the z axis). This suggests that when a deuterium molecule reorients itself to one of the other eleven alternative positions it is more likely to move to one of the positions where it lies above or below the xy plane rather than to position 2, 3 or 4 where it would be in the plane.

A final point of interest in these measurements is to consider the values obtained for the length of the D–Cl bond at various temperatures. At first sight, as the third column of *Table 4.6* shows, there is a substantial reduction of the bond length with increase of temperature. However, correction must be made for the apparent shortening of the bond due to the rotation of the D atom about the Cl position. The correction increases with the value of the angular rotations, and, as *Figure 4.15* shows, $\langle \phi_{in}^2 \rangle$ increases substantially above a temperature of 80 K. When the correction is made, as the later columns of *Table 4.6* show, there remains only a small change of bond length with rise of temperature. This slight increase may be significant and can be attributed to the relative strengthening of the *inter*molecular bond at low temperatures.

Later measurements by Niimura et al.[35] with single crystals have substantially confirmed these conclusions for DCl. The same group of workers[36] has examined single crystals of HCl itself and has concluded that there are two transitions in this case, with an intervening region

58 Structural Studies

Table 4.6 Variation of D–Cl bond length with temperature

T/K	Phase	Bond length/Å		
		uncorrected	correction	corrected value
4.2	orthorhombic	1.28 ± 0.02	0.02	1.30 ± 0.02
77	orthorhombic	1.28 ± 0.02	0.02	1.30 ± 0.02
92	orthorhombic	1.26 ± 0.02	0.03	1.29 ± 0.02
111	cubic	1.19 ± 0.03	0.09	1.28 ± 0.03

between the ferroelectric transition at 98.4 K and about 120 K for which a twinned orthorhombic structure exists. There remains some controversy about the nature of this phase.

Solid deuterium bromide occurs in three phases, with transitions at 93.5 and 120.3 K and the lower two phases have been examined, using powdered materials, by Sándor and Johnson[37,38]. In each case the neutron intensities, which are substantially dependent on the deuterium positions, revealed that the space group had not been correctly assessed from earlier X-ray data. The lower phase was found to be isomorphous with the low-temperature phase of DCl, being composed of parallel plane zig-zag chains in space group $Bb2_1m$. The 'middle phase', between the two transitions, was found to be disordered and each molecule had two possible orientations which were respectively the same as and the opposite of the orientation in the low-temperature ordered phase.

Deuterium sulphide likewise exists in three phases. Below 103.5 K there is an ordered phase (Sándor and Ogunade[39]), of space group $P4_2$ which is shown in projection in *Figure 4.16*. Each sulphur atom forms hydrogen bonds with two neighbouring sulphur atoms, one at a higher and one at a lower level than its own height in the unit cell, with the result that the hydrogen bonds form zig-zag chains parallel to the a and b axes. At 102 K there is a transition to a cubic structure of space group $Pa3$ in which there is orientational disorder of the molecules, each having six possible directions, leading to a continuously varying distribution of hydrogen bonds. More recent work, with improved resolution, suggests that the symmetry of the low-temperature phase is no higher than monoclinic.

Solid acetylene exists in two phases with a transition at 149 K. The low-temperature phase has been studied at liquid-helium temperature as a powder by Koski and Sándor[40]. The space-group is $Acam$ and the molecules lie parallel to the xy plane at levels of 0 and ½ as indicated in *Figure 4.17*. The accuracy of the powder refinement was not sufficient to determine anisotropic temperature factors and thus to assess the likely libration of the molecule. Accordingly the uncorrected value of 1.18 Å found for the C≡C bond length is likely to be considerably in error because of the shortening effect due to the libration.

An examination of solid deuterium, both above and below the λ transition, by Mucker *et al.*[41] is of particular interest because it underlines the way in which conclusions from powder diffraction measurements may be invalidated by preferred orientation in the samples. Much earlier work,

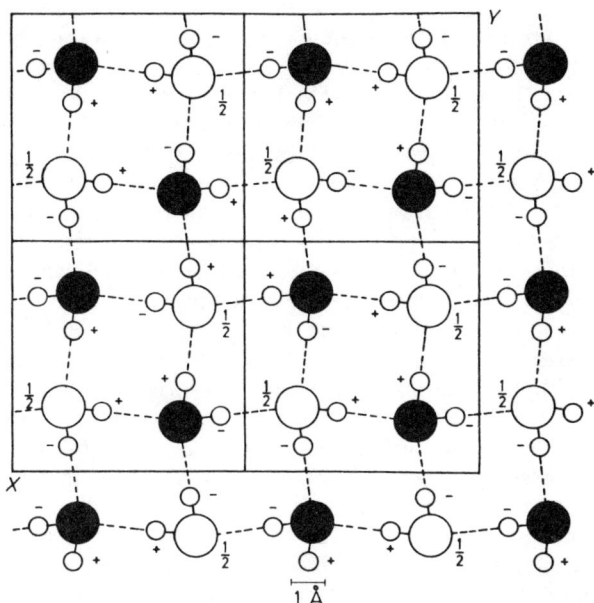

Figure 4.16 The hydrogen bonds in the structure of the ordered low-temperature phase of deuterium sulphide, D_2S, projected on the 001 plane. The large circles indicate sulphur atoms, either in the plane of the paper (filled circles) or at a height of c/2 (open circles). The small circles indicate deuterium atoms which link each sulphur atom via hydrogen bonds to two neighbouring sulphur atoms (From Sandor and Ogunade[39], by courtesy of Nature)

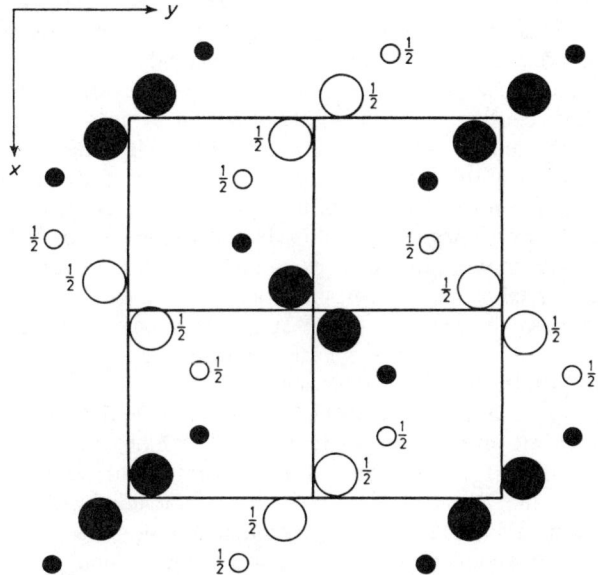

Figure 4.17 The structure of the low-temperature orthorhombic phase of acetylene. The large circles indicate carbon atoms and the small circles are hydrogen. The molecules formed by filled circles lie in the plane of the paper and those with open circles are at a height of c/2 (From Koski and Sandor[40], by courtesy of Acta Cryst.)

for example by Ozerov et al.[42], drew erroneous conclusions because of inaccurate intensity data caused by the difficulty of growing a randomly oriented powder. In the work of Mucker et al.[41] the deuterium was solidified in a porous sponge of aluminium-magnesium alloy which inhibited the growth of large crystals by limiting the regions containing the bulk liquid to small randomly-oriented pores and surfaces. Thus the usual tendency for crystal-growth with preferred orientation to occur on walls, with the 0002 plane parallel to the wall in the case of deuterium, was put to advantage. In fact the preferred orientation was reduced to 8%. Measurements were made at various temperatures in the range 1.9–13 K for samples containing para concentrations of 3, 33, 63 and 80%. In all cases solidifcation from the liquid produced a hexagonal structure in which the centres of the molecules (two per unit cell) were at the positions corresponding to hexagonal close packing. However, the structure is disordered and, when a correction is made for the residual orientation, the best agreement is obtained for a model in which the molecules precess about the z axis of the unit cell, with a precession angle which increases from 65° to 85° with the para concentration. For the sample containing 80% paradeuterium, transition to a cubic phase takes place below the λ point observed in the heat capacity. This cubic structure corresponds to space group $Pa3$ and has four molecules per unit cell. The molecule at the origin on the unit cell lies along the body diagonal of the cell and the remaining three molecules, with their centres located at the centres of the faces, lie along the remaining three directions of the $\{111\}$ form.

Solid oxygen has been much studied with neutrons, being of particular interest because it is the simplest of the few known materials whose magnetic properties arise from unpaired p electrons. Moreover, it is molecular, and not atomic, orbitals which give rise to the magnetism. There are three phases. On cooling, oxygen solidifies at 54 K into the cubic γ phase which transforms at 44 K to a rhombohedral β phase and, subsequently, at 24 K to a monoclinic α phase. Some of the earlier work was invalidated by samples being mixtures of both the α and β phases. The α phase is antiferromagnetic (Collins[43], Alikhanov et al.[44,45]) and the latter workers have shown that the magnetic moments are directed along the monoclinic crystal axis and perpendicular to the axis of the molecule. The ordered magnetic reflections (001) and (110) have intensities which are consistent with a molecular magnetic moment of 2 μ_B and the expected sharp form-factor for the extended orbitals. It is noteworthy that KO_2 was also examined by Alikhanov et al.[45] and below 7 K was found to exist as an antiferromagnetic phase in which the KO_2^- ion, with a moment of 1 μ_B, has a very similar form-factor to that for the O_2 molecule. In the β phase of oxygen it is concluded (Collins[43]) that the magnetic moments are not randomly aligned but show short-range order. For both α and β phases the measurements which we have described were carried out with powder samples. For the γ phase, X-ray measurements (Jordan et al.[46]) with single crystals identified the space group as $Pm3n$, with a high degree of rotational disorder among the molecules. There were found to be two types of oxygen molecule:

one, centred at the origin, showing spherical rotation and the other with a very anisotropic distribution equivalent to statistical disorder. This model was refined considerably further by unpolarized neutron observations with single crystals (Cox, Samuelsen and Beckurts[47]), it having been found that large crystals with a diameter of 1.5 cm could readily be grown. These measurements showed that the first type of O_2 molecules are in the 2(a) positions with their lengths randomly oriented along $\langle 111 \rangle$ axes; the remaining six molecules lie at the 6(d) positions with their lengths along $\langle 100 \rangle$ axes. In each case there is a large librational motion, with an amplitude of about 20°, superimposed on a substantial isotropic thermal motion. The libration produces a substantial apparent shortening of the O–O bond length. These conclusions have been strengthened by observations with a polarized neutron beam, with a field of 80 kOe applied to the crystals. By examining crystals oriented in various directions it was possible to measure the form-factor for magnetic scattering and its dependence on direction; the form-factor of each molecule will depend on the orientation of the molecule relative to the direction of the scattering vector. The measurements are in good agreement with the calculations of Leoni and Sacchetti[48].

Finally we note the relatively simple rare-gas compound XeO_2F_2 as a good example of the use of neutrons for crystallographic studies. For X-rays, of course, xenon is a heavy atom, with a scattering length of 15.3×10^{-12} cm, compared with 2.2 and 2.5×10^{-12} cm for oxygen and fluorine, respectively. For neutrons, on the other hand, the three atoms have scattering lengths which are closely similar, the values of b being 0.48, 0.57 and 0.56×10^{-12} cm for Xe, O and F, respectively. This results in a favourable situation for determining accurately the shape of the molecule by neutron diffraction. This has been done by Peterson, Willett and Huston[49] using small crystals measuring approximately $0.9 \times 0.8 \times 0.5$ mm, bathed in a neutron flux of 2.9×10^6 cm^{-2} s^{-1} at a wavelength of 1.14 Å. *Figure 4.18* shows the molecule and its

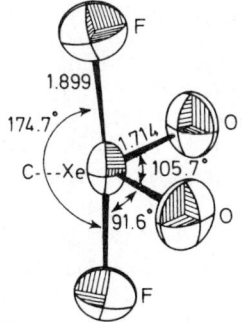

Figure 4.18 The molecule of XeO_2F_2, indicating the bond lengths and angles; it is essentially a trigonal bipyramid for which the third position C in the equatorial plane is not occupied by an atom, but is likely to be filled by a lone pair of electrons. The ellipsoids indicate the thermal motion of the atoms (From Peterson, Willett and Huston[49], by courtesy of J. Chem. Phys.)

62 *Structural Studies*

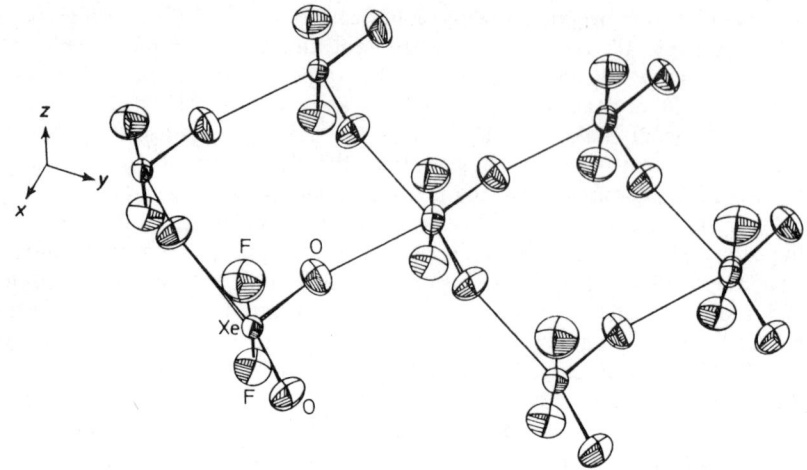

Figure 4.19 The layer structure of XeO_2F_2 in which the individual molecules are linked by weak Xe···O bridges (From Peterson, Willett and Huston[49], by courtesy of J. Chem. Phys.)

associated thermal ellipsoids. It may be described as a trigonal bipyramid, with the xenon atom at the centre and the fluorine atoms on the vertical axis, the oxygen atoms lie in the equatorial plane, occupying two of the three trigonally arranged sites and the third site is likely to be occupied by a lone-pair of electrons. *Figure 4.19* indicates the layer structure of XeO_2F_2 and it is seen that each oxygen atom is bridged to a xenon atom in a neighbouring molecule, giving an Xe—O distance of 2.81 Å, which is much shorter than the sum of the van der Waals' radii of xenon and oxygen (~3.6 Å). In contrast to these relatively rigid layers in the xy plane of the structure, successive layers are held together only by van der Waals' forces.

4.6 Amino-acid derivatives

The original neutron diffraction studies of hydrogen bonds were mainly concerned with compounds such as potassium dihydrogen phosphate and inorganic salts containing water of crystallization. These had relatively simple structures and a small number of parameters, offering the possibility of finding the positions of the hydrogen atoms, and assessing their thermal motion, under the conditions of relatively low accuracy and resolution provided at the early nuclear reactors. As more intense neutron beams have become available the accuracy of determination has increased immensely and, accordingly, much more complicated structures have proved susceptible to study. The scope of recent work can be well illustrated by considering some of the studies of amino-acids and their derivatives which have been carried out at the Brookhaven National Laboratory, mainly under the guidance of the late W.C. Hamilton. These studies of some of the units which build up proteins and nucleic acids are important because they give important information about the geometry of the framework of these macromolecules, bearing in mind that the secondary and tertiary structures of polypeptide and polynucleotide chains depend on hydrogen bonds and van der Waals' contacts between hydrogen atoms.

As examples of this work we shall first consider and compare the results for

(a) the amino-acid L-alanine

$$\begin{array}{c} CH_3CHCOO^- \\ | \\ NH_3^+ \end{array}$$

(Lehmann, Koetzle and Hamilton[50])

(b) the monohydrate of the monoamino-acid L-asparagine

$$\begin{array}{c} CONH_2 \\ | \\ CH_2CHCOO^- \\ | \\ NH_3^+ \end{array}$$

(Verbist et al.[51])
and (c) the dihydrate of the diamino-acid L-arginine

$$\begin{array}{c} CH_2CH_2NHC(NH_2^+)NH_2 \\ | \\ CH_2CH(NH_2)COO^- \end{array}$$

(Lehmann et al.[52])

For each of these molecules the projections of scattering amplitude show clearly the nature of the zwitterion, with transfer of a hydrogen atom from the carboxyl group to an NH or NH_2 group, giving separation of negative and positive charges. The relations between the three molecules are illustrated best by the diagrams of their shapes revealed by the neutron diffraction measurements. These are shown in *Figure 4.20*, where the ellipsoids indicate the anisotropic thermal motion of the atoms which we shall discuss further later on. In each case all the atomic positions and the thermal motions of the atoms have been determined, to an accuracy indicated by an R factor of 2.2, 2.6 and 3.4%, respectively. The linear dimensions of the crystals ranged from 2 to about 3 mm and 813, 1003 and 1627 independent reflections were measured in the three cases. In general the non-hydrogen atoms were located with an accuracy 0.001–0.002 Å and the hydrogen atoms to 0.002–0.003 Å, thus enabling the details of the hydrogen bonds to be determined with precision.

For L-alanine it is found that there are three hydrogen bonds, indicated in *Figure 4.20(a)*, which link the hydrogen atoms of the a-NH_3^+ group to oxygen atoms in adjacent molecules. One set of these bonds

64 *Structural Studies*

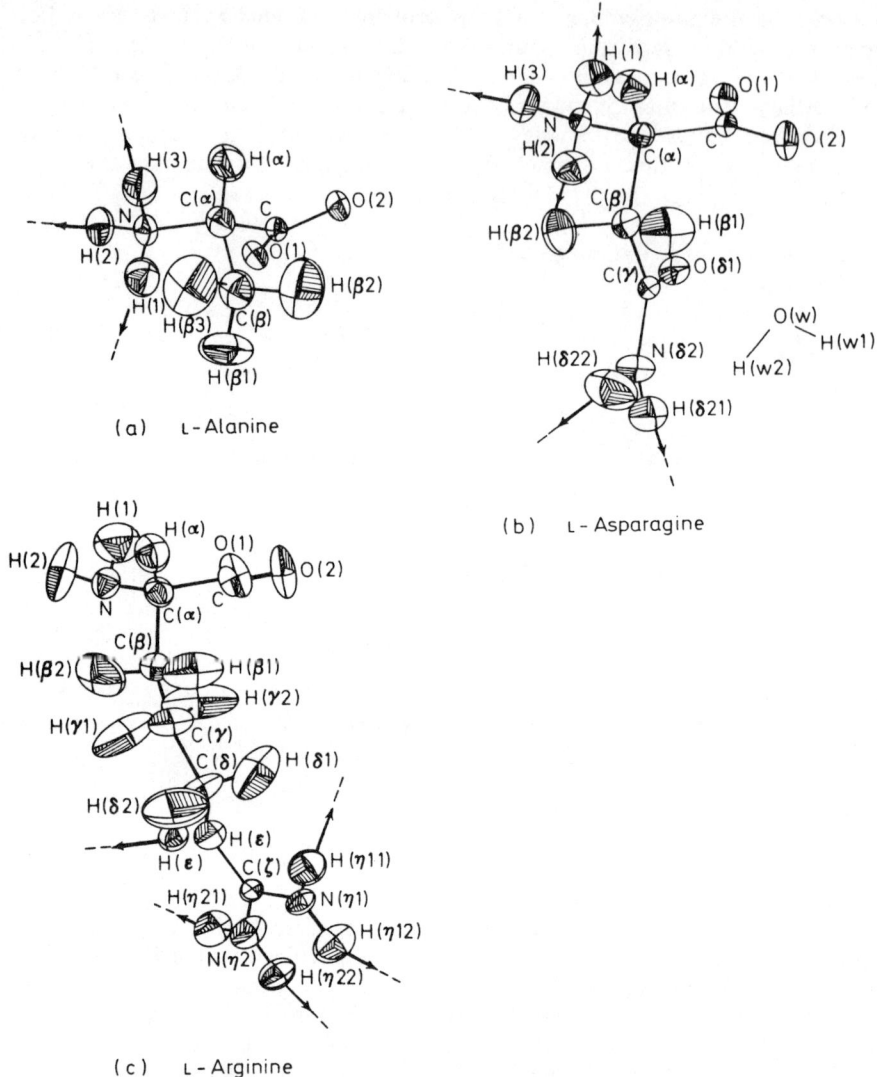

Figure 4.20 The molecules of (a) L-alanine, (b) L-asparagine and (c) L-arginine as revealed by neutron diffraction. The ellipsoids indicate the thermal motion of the atoms. The arrows indicate intermolecular hydrogen bonds (From the papers of Lehmann et al.[50], Verbist et al.[51] and Lehmann et al.[52])

link the molecules together to form chains along the c-axis, whereas the other two bind together the chains to make a three-dimensional network. In asparagine monohydrate there are seven different intermolecular hydrogen bonds and, again (see *Figure 4.20(b)*), three of these link the a-NH$_3^+$ group to adjacent oxygens; two others link the NH$_2$ group and a further two link the water molecule. On the other hand, in L-arginine dihydrate the a-NH$_2$ group, which here is neutral, does not take part in hydrogen bonding and, of the nine hydrogen bonds, five originate in the guanidinium group and four join the hydrogen atoms in the two water molecules.

In each of the above cases the thermal motion has been assessed in terms of partially rigid bodies in which the main framework of the molecule moves as a whole but groups such as NH_3 and CH_3 have independent, albeit hindered, rotation. The type of information which has been obtained can be well illustrated by comparing the conclusions for the motion of the a-NH_3^+ or NH_2 group which appears at the top left-hand corner of each of the three molecules in *Figure 4.20*. For L-alanine the root-mean-square amplitude of libration was about 6°, corresponding to a barrier to rotation of 20 kcal mol^{-1}, in contrast to a rotation of 9° and 5.6 kcal mol^{-1} for the methyl group. In L-asparagine, where the H$\cdots$O distances are larger, the hydrogen bonds are weaker and there is a libration of 8½° for the NH_3^+ group, indicating a reduced barrier to rotation of 7.1 kcal mol^{-1}. More strikingly, in L-arginine, where the NH_2 group is not hydrogen-bonded, its libration has increased to about 17° indicating a very low barrier of only 1.6 kcal mol^{-1}. On the other hand, in this same molecule the two NH_2 groups within the guanidinium group, which take part in the hydrogen bonds, have librations of only 6° and 7°, respectively.

Among other papers describing this series of investigations of amino-acid derivatives is an account of a study of L-lysine monohydrochloride dihydrate by Koetzle et al.[53] and there is a concurrent study by Bugayong, Sequeira and Chidambaram[54]. A statistical comparison of the two sets of conclusions is provided in the second of these papers and gives useful information on the accuracies which can be expected in this type of measurement. Two other topics of general interest receive instructive treatment in these papers. First, the correction of bond lengths for the shortening produced by anisotropic thermal motion, which we mentioned on p. 41, is discussed at some length, particularly for L-lysine. Secondly, because these neutron studies were preceded by very accurate X-ray work in which hydrogen positions were determined, there are some good examples of the distinction which exists between bond lengths such as N–H and O–H when measured in turn by X-rays and neutrons, because the two radiations measure respectively the position of the centroid of the electron cloud and the position of the nucleus. *Table 4.7* gives some comparative figures for bonds in L-asparagine monohydrate, with the nomenclature of *Figure 4.20(b)*: O(w), H(w1) and H(w2) are the atoms of the water molecule. Later publications in this same series of papers describe studies of L-tyrosine and L-tyrosine hydrochloride (Frey et al.[55]), L-serine (Frey et al.[56]), to which we shall refer again in the following chapter in our discussion of 'direct methods' of analysis, and a study of the hydrogen bonding in the purine-pyrimidine base pair 9-methyladenine-1-methylthymine (Frey et al.[57]). The latter is of particular interest in demonstrating the existence of disorder in the structure, which neutrons are able to distinguish because of their ability to differentiate between nitrogen and carbon atoms, which are not distinguishable with X-rays. About 90% of the adenine–thymine pairs have the so-called Hoogsteen configuration but the remainder have the reversed configuration

Structural Studies

Table 4.7 Differences between X-ray and neutron bond lengths

Bond	Length/Å	
	X-ray	Neutron
N–H(1)	0.88	1.053 (2)
N–H(2)	0.90	1.044 (2)
N–H(3)	0.87	1.053 (2)
N($\delta 2$)–H($\delta 21$)	0.78	1.033 (2)
N($\delta 2$)–H($\delta 22$)	0.93	1.015 (3)
O(w)–H(w1)	0.80	0.969 (3)
O(w)–H(w2)	0.78	0.963 (3)

which is achieved by a rotation of the thymine molecule by 180° about N(3)···C(6), which effectively interchanges the nitrogen atom N(1) and the carbon atom C(5). This conclusion rests on the fact that the refined values of the neutron scattering lengths for N(1) and C(5) are 0.884 and 0.691 $\times$ 10^{-12} cm, respectively; these values are significantly different from 0.910 and 0.657 $\times$ 10^{-12} cm which are the average values for the remaining nitrogen and carbon atoms in the molecule.

References

1. BACON, G.E., *Proc. Roy. Soc.*, **A209**, 397 (1951); BACON, G.E. and LOWDE, R.D., *Acta Cryst.*, **1**, 303 (1948)
2. PETERSON, S.W. and LEVY, H.A., *J. Chem. Phys.*, **19**, 1416 (1951)
3. BROWN, G.M. and LEVY, H.A., *Acta Cryst.*, **B29**, 790 (1973)
4. BROWN, G.M. and LEVY, H.A., *Science*, **141**, 921 (1963)
5. BUSING, W.R. and LEVY, H.A., *J. Chem. Phys.*, **26**, 563 (1957)
6. BACON, G.E., *Neutron Diffraction*, 3rd edn., 363, Clarendon Press, Oxford (1975)
7. PETERSON, S.W. and LEVY, H.A., *J. Chem. Phys.*, **20**, 704 (1952)
8. MCGAW, G.L. and IBERS, J.A., *J. Chem. Phys.*, **39**, 2677 (1963)
9. MANOJLOVIC, Lj. and SPEAKMAN, J.C., *Acta Cryst.*, **B24**, 323 (1968)
10. BACON, G.E., WALKER, C.R. and SPEAKMAN, J.C., *J. Chem. Soc. Perkin Trans. II*, in press
11. WILLIAMS, J.M., PETERSON, S.W. and LEVY, H.A., *American Crystallographic Association, Program Abstracts I7*, Albuquerque, N.M. (1972)
12. WILLIAMS, J.M. and SCHNEEMEYER, L.F., *J. Amer. Chem. Soc.*, **95**, 5780 (1973)
13. LUNDGREN, J.-O. and WILLIAMS, J.M., *J. Chem. Phys.*, **58**, 788 (1973)
14. FINHOLT, J.E. and WILLIAMS, J.M., *J. Chem. Phys.*, **59**, 5114 (1973)
15. HAMILTON, W.C. and IBERS, J.A., *Acta Cryst.*, **16**, 1209 (1963)
16. DELAPLANE, R.G., IBERS, J.A., FERRARO, J.R. and RUSH, J.J., *J. Chem. Phys.*, **50**, 1920 (1969)
17. KAY, M.I., OKAYA, Y. and COX, D.E., *Acta Cryst.*, **B27**, 26 (1971)
18. ERMER, O. and DUNITZ, J.D., *Chem. Commun.*, 178 (1971)
19. BREGMAN, J., HIRSHFELD, F.L., RABINOVICH, D. and SCHMIDT, G.M.J., *Acta Cryst.*, **19**, 227 (1965)

20. HIRSHFELD, F.L. and RABINOVICH, D., *Acta Cryst.*, **19**, 235 (1965)
21. AURIVILLIUS, K. and STÅLHANDSKE, C., *Z. Krist.*, **142**, 129 (1975)
22. AURIVILLIUS, K., *Acta Chem. Scand.*, **26**, 2113 (1972)
23. MASAKI, N. and DOI, K., *Acta Cryst.*, **B28**, 785 (1972)
24. SILLEN, L.G. and NYLANDER, A.L., *Arkiv. Kemi. Miner. Geol.*, **17A**, No.4 (1943)
25. ZALKIN, A. and TEMPLETON, D.H., *J. Chem. Phys.*, **40**, 501 (1964)
26. KAY, M.I., FRAZER, B.C. and ALMODOVAR, I., *J. Chem. Phys.*, **40**, 504 (1964)
27. LOOPSTRA, B.O. and RIETVELD, H.M., *Acta Cryst.*, **B25**, 787 (1969)
28. TAYLOR, J.C. and WILSON, P.W., *J. Solid State Chem.*, **14**, 378 (1975)
29. LEVY, J.H., TAYLOR, J.C. and WILSON, P.W., *Acta Cryst.*, **B31**, 880 (1975)
30. LECIEJEWICZ, J., MURASIK, A. and TROC, R., *Phys. Status Solidi*, **30**, 157 (1968)
31. LECIEJEWICZ, J., MURASIK, A., TROC, R. and PALEWSKI, T., *Phys. Status Solidi (b)*, **46**, 391 (1971); *Phys. Status Solidi (b)*, **48**, 445 (1971)
32. SÁNDOR, E. and FARROW, R.F.C., *Nature*, **213**, 171 (1967)
33. SÁNDOR, E. and FARROW, R.F.C., *Nature*, **215**, 1265 (1967)
34. SÁNDOR, E. and FARROW, R.F.C., *Discuss. Faraday Soc.*, **48**, 78 (1969)
35. NIIMURA, N., FUJII, Y., MOTEGI, H. and HOSHINO, S., *J. Phys. Soc. Japan*, **35**, 842 (1973)
36. NIIMURA, N., SHIMAOKA, K., MOTEGI, H. and HOSHINO, S., *J. Phys. Soc. Japan*, **32**, 1019 (1972)
37. SANDOR, E. and JOHNSON, M.W., *Nature*, **217**, 541 (1968)
38. SÁNDOR, E. and JOHNSON, M.W., *Nature*, **223**, 730 (1969)
39. SÁNDOR, E. and OGUNADE, S.O., *Nature*, **224**, 905 (1969)
40. KOSKI, H.K. and SÁNDOR, E., *Acta Cryst.*, **B31**, 350 (1975)
41. MUCKER, H.F., HARRIS, P.M., WHITE, D. and ERICKSON, R.A., *J. Chem. Phys.*, **49**, 1922 (1968)
42. OZEROV, R.P., KOGAN, V.S., ZHDANOV, G.S. and LAZAREV, B.G., *J. Phys. Soc. Japan*, **27**, Suppl. BII, 385 (1962)
43. COLLINS, M.F., *Proc. Phys. Soc.*, **89**, 415 (1966)
44. ALIKHANOV, R.A., *JETP Lett.*, **5**, 349 (1967)
45. ALIKHANOV, R.A., ILYINA, I.L. and SMIRNOV, L.S., *Phys. Status Solidi (b)*, **50**, 385 (1972)
46. JORDAN, T.H., STREIB, W.E., SMITH, H.W. and LIPSCOMB, W.N., *Acta Cryst.*, **17**, 777 (1964)
47. COX, D.E., SAMUELSON, E.J. and BECKURTS, K.H., *Phys. Rev. B*, **7**, 3102 (1973)
48. LEONI, F. and SACCHETTI, F., *Phys. Rev. B*, **7**, 3122 (1973)
49. PETERSON, S.W., WILLETT, R.D. and HUSTON, J.L., *J. Chem. Phys.*, **59**, 453 (1973)
50. LEHMANN, M.S., KOETZLE, T.F. and HAMILTON, W.C., *J. Amer. Chem. Soc.*, **94**, 2657 (1972)
51. VERBIST, J.J., LEHMANN, M.S., KOETZLE, T.F. and HAMILTON, W.C., *Acta Cryst.*, **B28**, 3006 (1972)
52. LEHMANN, M.S., VERBIST, J.J., HAMILTON, W.C. and KOETZLE, T.F., *J. Chem. Soc. Perkin Trans. II*, 133 (1973)

53. KOETZLE, T.F., LEHMANN, M.S., VERBIST, J.J. and HAMILTON, W.C., *Acta Cryst.*, **B28**, 3207 (1972)
54. BUGAYONG, R.R., SEQUEIRA, A. and CHIDAMBARAM, R., *Acta Cryst.*, **B28**, 3214 (1972)
55. FREY, M.N., KOETZLE, T.F., LEHMANN, M.S. and HAMILTON, W.C., *J. Chem. Phys.*, **58**, 2547 (1973)
56. FREY, M.N., LEHMANN, M.S., KOETZLE, T.F. and HAMILTON, W.C., *Acta Cryst.*, **B29**, 876 (1973)
57. FREY, M.N., KOETZLE, T.F., LEHMANN, M.S. and HAMILTON, W.C., *J. Chem. Phys.*, **59**, 915 (1973)
58. WOLLAN, E.O., DAVIDSON, W.L. and SHULL, C.G., *Phys. Rev.*, **75**, 1348 (1949)
59. PETERSON, S.W. and LEVY, H.A., *Acta Cryst.*, **10**, 70 (1957)
60. KAMB, B., HAMILTON, W.C., LAPLACA, S.J. and PRAKASH, A., *J. Chem. Phys.*, **55**, 1934 (1971)
61. ARNOLD, G.P., WENTZER, R.G., RABIDEAN, S.W., NERESEN, N.G. and BOWMAN, R.L., *J. Chem. Phys.*, **55**, 589 (1971)
62. SPEAKMAN, J.C., in *Molecular Structure by Diffraction Methods*, Vol.I, 208, The Chemical Society, London (1973)
63. BRUNTON, G.D. and JOHNSON, C.K., *J. Chem. Phys.*, **62**, 3797 (1975)

5

DIRECT METHODS OF STRUCTURE ANALYSIS

Recent years have shown many advances in the theory underlying 'direct methods' of phase determination in the process of structure analysis. Together with concurrent development of computer programs for most of the operations this has led to widespread use of these methods. 'Direct methods' depend on the fact that structure factors, and their phases, are mutually related and not independent of each other, so that a wide range of inequalities and probabilities exist among the structure factors, dependent on correlations among their indices. Many of these correlations depend on two principles: first, the scattering power in solids is concentrated into atoms, which are regions of characteristic shape and size, and, secondly, in the case of X-ray scattering all the scattering matter has the same sign, being due to negative electrons.

In the case of neutron diffraction there is, in principle at least, a complication in the application of direct methods arising from the fact that scattering lengths of atoms are *not* all of the same sign: hydrogen, for example, has a negative scattering length. Sikka[1] estimated that the negative sign would be a limitation in any material for which the proportion of scattering due to the hydrogen atoms, and defined as

$$\sum_{\text{hydrogen}} b_\text{H}^2 \Big/ \sum_{\text{all}} b^2$$

is greater than about 25%. Bearing in mind the scattering amplitudes of the various atoms in the Periodic Table, it is found that this limit would permit 50 atomic per cent of hydrogen in most materials and is not a very severe restriction. Indeed, Verbist *et al.*[2] conclude that direct methods can probably be applied to neutron data for any crystal for which they are applicable to X-ray data.

Any difficulty with hydrogen can of course be avoided by using deuterated material and this may also be of advantage in interpreting Patterson functions. In a neutron Patterson map, peaks between pairs of hydrogen atoms will be positive but peaks between hydrogen and some other atom will normally be negative. This can be a help in the identification of such peaks, but will be a drawback in situations where there are overlapping peaks and the overlap of positive and negative peaks could lead to the complete disappearance of these peaks. In these circumstances it is very advantageous to compare the Patterson maps obtained with normal and deuterated material.

At the same time, neutron data are more amenable to analysis by those direct methods, such as the use of Sayre's equation, which depend on an assumption that all the atoms within a molecule are of the same scattering power. With a few exceptions this is generally a good approximation for neutrons, although it is certainly not often so for X-rays. This circumstance counter-balances the fact that the 'heavy atom' technique, much used with X-ray data, is not often applicable to neutron results.

As an early example of the use of direct methods we may first note the study of trichloracetic acid by Jonsson and Hamilton[3]. This compound contains only one atom of hydrogen in a molecule of eight atoms and its structure was successfully solved by application of the Σ_2 sign-determining relationship of Karle and Karle[4]. The superacid salt potassium trihydrogen disuccinate $KH_3(CO_2CH_2CH_2CO_2)_2$ has proved to be a favourable case, and has been solved directly from neutron data by Dunlop and Speakman[5]. In this molecule about 20% of the scattering is due to hydrogen and the atomic percentage of the latter is about 40%. 224 normalized structure amplitudes having $E > 1.5$ were used in solving the phase problem after positive signs had been allocated to the three reflections 621, 51$\bar{7}$ and 43$\bar{8}$ to fix the origin. The program PHASE of the X-RAY SYSTEM then generated a set of signs for 206 of the data. The Fourier map of scattering density based on these terms revealed every atom of the structure and it was found subsequently that all but 11 of these allocated signs were correct.

An alternative approach to overcome the difficulty caused by the negative scattering density has been developed by Karle[6] and others, known as the 'squared structure' approach. This considers a hypothetical structure for which the atomic scattering amplitudes are the squares, b_r^2, of those, b_r, which exist in the real structure; accordingly all these 'new' values are positive. Probability relations are used to estimate the structure amplitudes of the squared structure from those measured experimentally for the real structure. The former are then employed in the sign-determining formulae to arrive at the signs for each reflection from the squared structure. This method was used by Ellison, Johnson and Levy[7] to solve directly the structure of glycollic acid from neutron data. This molecule, $CH_2OHCOOH$, contains 45 atomic percent of hydrogen, contributing 23% of the scattering, and is of particular interest because the neutron Patterson map has important missing peaks on account of the cancellation of positive and negative contributions in the manner which we discussed earlier. This circumstance had accounted for the failure of an earlier attempt to solve the structure of glycollic acid from neutron data using a method of Patterson analysis which had proved successful for potassium hydrogen chloromaleate, $HOOCCCl=CHCOO^-K^+$ (Ellison and Levy[8]).

The examples which we have discussed so far are all centrosymmetric structures and, accordingly, the process of phase determination for the reflections merely means a distinction between + and −. We mention three examples of structures without a centre of symmetry which have been determined from neutron data by direct methods. Verbist et al.[2] have examined L-proline monohydrate, an amino-acid which has the formula $C_5H_{11}O_3N$, including a five-membered ring, and which belongs to space group C2. There is 55 atomic percent of hydrogen, accounting for 28 percent of the scattering power, so that we are close to the limit suggested by Sikka[1]. 664 independent reflections were measured and 210 of these were used to generate 1440 Σ_2 relationships for the symbolic-addition method of Karle and Karle[4]. Five starting phases were then selected, two to define the origin, and then, in turn, the 32 possible combinations of phase value 45°, 135°, 225° and 315° for three other reflections. The weighted tangent-refinement method of Germain, Main and Woolfson[9] then produced 32 sets of refined phases for the 210 reflections. Examination of the solutions with high figures of merit led to a well-resolved picture of atomic peaks which was subsequently refined by structure factor calculations and Fourier syntheses.

Similar methods have been followed by Bernal and Watkins[10] in a study of the natural product melampodin, which has a large complex molecule $C_{21}H_{24}O_9$ and a non-centrosymmetric structure with space group $P2_12_12_1$ and 216 atoms in the unit cell. 414 reflections were used in the initial determination of phases, resulting in an E map from which 16 of the 21 carbon atoms and 5 of the 9 oxygen atoms were located. Subsequent least-squares analysis of the intensity data for 2303 independent neutron reflections then yielded an R factor of 5%. The particular interest of this determination arises from the large number of atoms in the unit cell. Sikka's study[1] had been restricted to cases where there were not more than 100 atoms in the unit cell.

More recently the structure of L-serine monohydrate, $CH_3C(OH)NH_2COOH \cdot H_2O$, has been determined by Frey et al.[11]. A similar procedure to that adopted for melampodin was used, employing the program MULTAN of Germain, Main and Woolfson[9]. The phases of 175

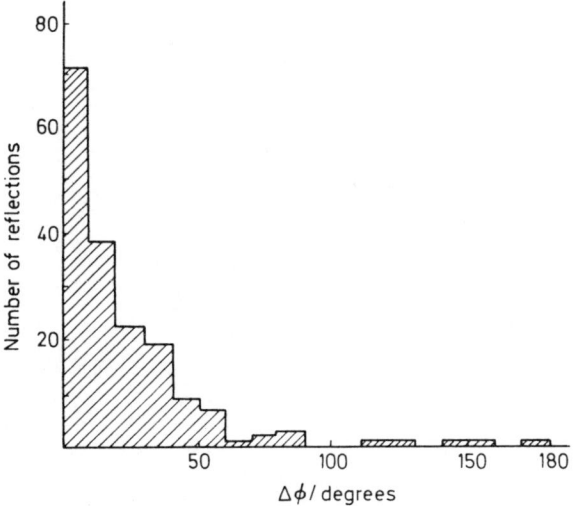

Figure 5.1 Histogram of the deviations, in degrees, between the phases determined by direct methods and those computed from the results of the final least-squares refinement, for L-serine monohydrate (From Frey et al.[11], by courtesy of Acta Cryst.)

reflections were determined by the weighted tangent formula and the average difference $\Delta\phi$ between these phases and those finally accepted from the least-squares analysis was 22°. The histogram of *Figure 5.1* shows how the values of the error were distributed among the 175 reflections.

References

1. SIKKA, S.K., *Acta Cryst.*, **A25,** 539 (1969)
2. VERBIST, J.J., LEHMANN, M.S., KOETZLE, T.F. and HAMILTON, W.C., *Nature*, **235,** 328 (1972)
3. JONSSON, P.-G. and HAMILTON, W.C., *J. Chem. Phys.*, **56,** 4433 (1972)
4. KARLE, J. and KARLE, I.L., *Acta Cryst.*, **21,** 849 (1966)
5. DUNLOP, R.S. and SPEAKMAN, J.C., *Z. Krist.*, **138,** 100 (1973)
6. KARLE, J., *Acta Cryst.*, **20,** 881 (1966)
7. ELLISON, R.D., JOHNSON, C.K. and LEVY, H.A., *Acta Cryst.*, **B27,** 333 (1971)
8. ELLISON, R.D. and LEVY, H.A., *Acta Cryst.*, **19,** 260 (1965)
9. GERMAIN, G., MAIN, P. and WOOLFSON, M.M., *Acta Cryst.*, **A27,** 368 (1971)
10. BERNAL, I. and WATKINS, S.F., *Science*, **178,** 1282 (1972)
11. FREY, M.N., LEHMANN, M.S., KOETZLE, T.F. and HAMILTON, W.C., *Acta Cryst.*, **B29,** 876 (1973)

6

CORRELATION OF X-RAY AND NEUTRON DATA: X–N SYNTHESES

In considering the advantages of using either neutrons or X-rays for studying some individual problem in physical or chemical crystallography, it has often been pointed out that the two radiations measure different quantities. More specifically, neutrons determine the positions of the atomic nuclei whereas X-rays determine the distribution of the surrounding electrons. Bearing in mind that atoms are not at rest but undergo thermal vibrations about their mean positions, this means that the time-averaged positions of the nuclei and the centre-of-gravity of the electrons are determined in the two cases. For an isolated atom these two positions would coincide, but this is no longer the case when the atom is part of a radical or molecule. In these cases the distribution of the electrons ceases to be spherical, or even centrosymmetric, and there is concentration of electron density within chemical bonds and also as 'lone pairs' within atoms. Bearing in mind the fact that the thermal motion is different for different atoms, and also anisotropic, it becomes evident that the electron distribution over a molecule may differ considerably in detail from the concept of a simple arrangement of spherical distributions of electrons.

The extent to which the intricate details of an electron distribution may be revealed in a diffraction analysis will depend on the accuracy with which the spectral intensities are measured. For simple arrangements of atoms it is possible to calculate the expected differences between electron and nuclear positions. These normally amount to a few hundredths of an ångstrom unit, but it is only within recent years that the accuracy with which the intensities of diffracted beams of neutrons and X-rays can be measured has become adequate to justify a precise comparison of atomic positions determined separately by the two radiations. We shall show that in appropriate cases the distinction between nuclear and electron distributions can be clearly demonstrated and interpreted. First, however, we emphasize that the *neutron* data can be interpreted straightforwardly. On the other hand, when conventional methods of least-squares analysis are applied to X-ray data the apparent accuracy of the model which emerges may be unduly optimistic. If a spherical model is used for the individual atoms, then the inadequacies of this model will result in false displacements and motions being accredited to the atoms in compensation; accordingly the discrepancy factor, R, achieved may not be realistic. It may be argued therefore that neutron diffraction should, in principle, be used first to determine atomic coordinates and thermal motion. X-ray diffraction can then be employed to determine the details of the electron distributions of the atoms, notably the way in which the electrons depart from simple spherical distributions.

A classical molecule for which full data from neutrons and X-rays were examined concurrently was hexamethylenetetramine (Duckworth, Willis and Pawley[1,2]) for which the neutron measurements were combined with the earlier X-ray data of Becka and Cruickshank[3]. In this work the procedure was to carry out a joint refinement in terms of two sets of positional parameters, provided separately by the neutron and X-ray data, and a single set of thermal parameters, utilizing of course two separate scale factors. In later studies with other materials it became customary to use the procedure which we have outlined above, whereby both coordinates and thermal parameters are deduced solely from the neutron data, in order that the X-ray data should play no part in determining the thermal parameters, thus avoiding any correlation between these and

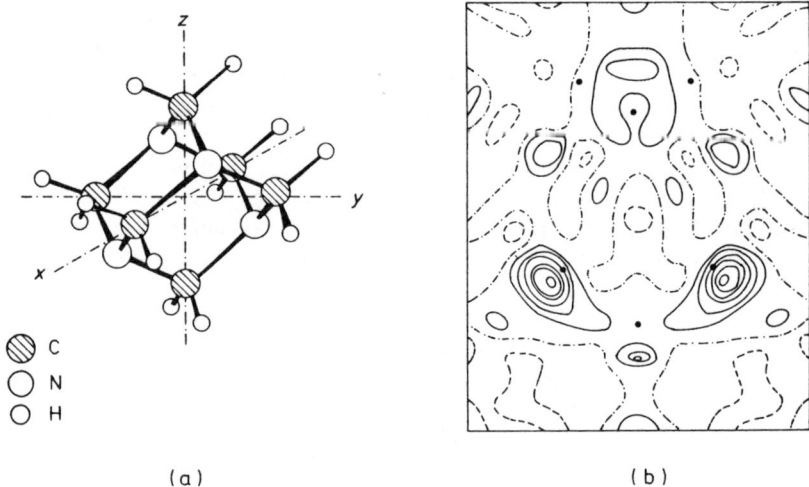

Figure 6.1 (a) shows the molecular structure of hexamethylenetetramine, $C_6N_4H_{12}$. (b) is a Fourier synthesis of $F_x - F'$, where F_x is the experimentally determined structure factor for X-rays and F' is a calculated value for X-rays using the position coordinates given by neutrons. The positions of the atomic nuclei in (b) are marked by the black dots and the large peak near each nitrogen atom indicates the lone pair electrons of the nitrogen (From Duckworth, Willis and Pawley[2], by courtesy of Acta Cryst.)

the inadequate spherical models for the electrons within an atom. Following their own procedure, Duckworth, Willis and Pawley[2] then plotted a Fourier map of the function $F_x - F'$, where F_x is the structure factor deduced from the observed X-ray intensity and F' is the calculated value for X-rays using the position coordinates given by neutrons. The map which was obtained as a (110) section appears in Figure 6.1, which also includes a diagram of the molecule and reveals very dramatically the lone-pair electrons of the nitrogen atoms, displaced from the nuclear positions in a direction away from the centre of the molecule. This follows because the function $F_x - F'$ will provide a map of the departure of the electron distribution from spherical symmetry. The precise value of the shift on this map will depend on series-termination effects. Coppens and Coulson[4] have calculated that the presence of the lone-pair electrons in nitrogen would displace the centre of gravity of the electron cloud by about 0.016 Å from the nitrogen nucleus and this is in good agreement with the difference of 0.018 Å between the atomic positions given by neutrons and X-rays. This displacement is about four times as large as the experimental standard deviation.

Another good example of the location of lone-pairs is provided by cyanuric acid, $C_3N_3H_3O_3$, studied by Coppens and Vos[5]. Both these measurements and the associated X-ray work of Verschoor and Keulen[6] were carried out near the temperature of liquid nitrogen in order to improve the accuracy of the high-angle data, to reduce the anharmonicity of the molecular vibrations and to reduce the thermal diffuse scattering. *Figure 6.2* is a difference plot of the function $F_x - F'$, where F_x is the structure factor measured with X-rays and F' is a calculated value of the structure factor for X-rays when the atomic coordinates determined with neutrons are used. In this case the thermal parameters measured with X-rays were used because of the difficulty of carrying out the X-ray and neutron experiments at the same temperature. The plot shows very clearly the two lone-pairs of the carbonyl oxygen atoms and the electrons concentrated at the centres of the bonds. For each of the oxygen atoms

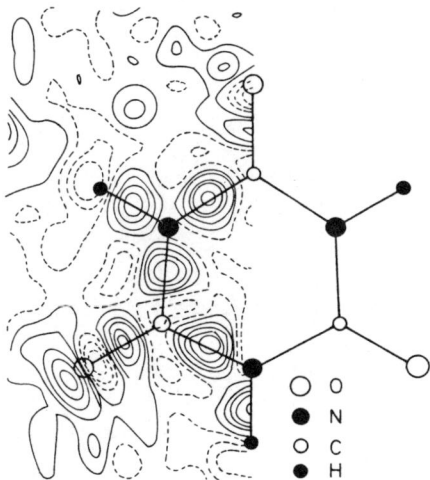

Figure 6.2 *A difference plot of electron density for cyanuric acid, using the function $F_x - F'$, where F_x is the measured structure factor for X-rays and F' is a calculated value for X-rays using atomic coordinates determined with neutrons. Zero and negative contours are shown as broken lines (From Coppens and Vos[5], by courtesy of Acta Cryst.)*

the peak, as measured with X-rays, is shifted towards the lone-pairs by 0.003 ± 0.001 and 0.006 ± 0.001 Å, whereas the two radiations give positions for the carbon and nitrogen atoms which are indistinguishable within the experimental accuracy.

The earliest example is *sym*-triazine, $C_3N_3H_3$ (Coppens[7]) for which *Figure 6.3* shows a difference plot of $F_x - F$, where F is a calculated structure factor for X-rays using the X-ray form-factor for a spherically-symmetric atom but both the atomic coordinates and thermal parameters deduced from the neutron measurements. The figure thus shows the departures of the electron density in the molecule from what would be expected for spherical atoms, i.e. the redistribution of charge density in going from isolated atoms to the molecule. The redistribution is revealed as the electron lone-pair behind the nitrogen atom and the excess of electrons near the centres of the C–N bonds, brought about by a withdrawal of electrons from the positions of the nuclei. The atoms show up as negative depressions since the spherical atom model has placed excess electron density in these regions, at the expense of bonding and lone-pair density which then appear in the difference map as regions of positive electron density. As we have emphasized earlier, the importance of using neutron thermal data in the calculation of F is to avoid a false

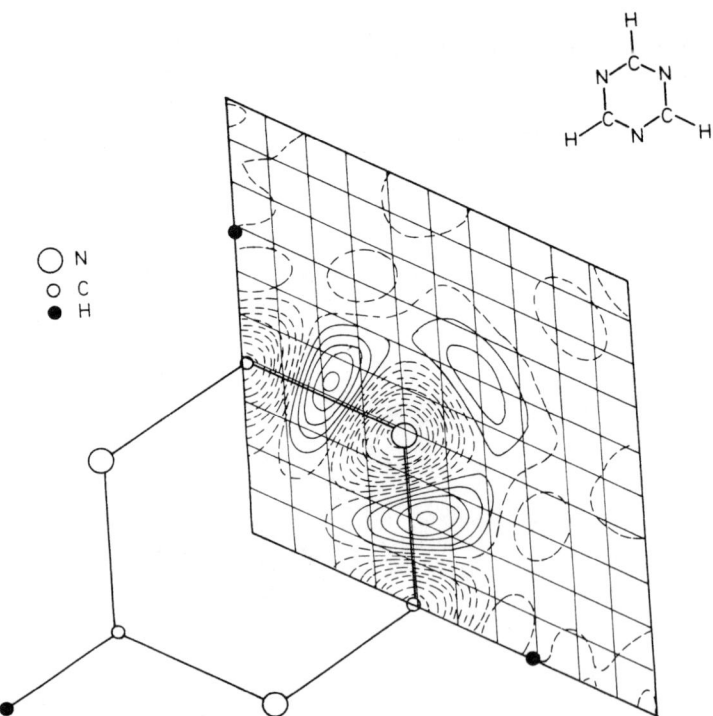

Figure 6.3 A difference map through the molecular plane of sym-triazine, with, inset, a sketch of the molecule (From Coppens[7], by courtesy of Science)

interpretation of the lone-pair density as an anisotropic thermal motion in the direction of the lone pair. A direct demonstration of this is provided by *Figure 6.4*, which shows the *difference* between the thermal ellipsoids as found by neutrons and X-rays; with X-rays it is concluded, falsely, that there is extra vibration of the nitrogen atom in the direction of the overlap density in the C—N bonds (Coppens[8]). A similar demonstration has been given by Coppens *et al.*[9] from neutron and X-ray data for α-deutero-oxalic acid dihydrate. The difference projection for this is shown in *Figure 6.5(a)*, and *6.5(b)* shows the difference ellipsoids. There is a marked concentration of electron density at the centre of the C—C bond and in both the two lone-pairs around the carbonyl oxygen and the single lone-pair behind the oxygen atom of the hydroxyl group. As a result, X-rays incorrectly suggest extra motion of the carbon atoms in the direction of the C—C bond and enhanced motion of the oxygen atoms in the directions of their lone-pairs.

A detailed account of the departures of the electron density in tetracyanoethylene oxide, $C_2(CN)_4O$, from the spherical model has been given by Matthews and Stucky[10], using neutron and X-ray data collected by Matthews *et al.*[11]. The latter emphasize the importance of accurate correction for extinction effects. Conclusions about any electrons which are concentrated at bond centres, remote from the atomic nuclei, will be very dependent on accurate intensity measurements at low angles of scattering, and it is the intense low-angle reflections for which the effect of extinction is most severe. *Figure 6.6* shows the molecular structure and the anisotropic thermal ellipsoids for tetracyanoethylene oxide, alongside the Fourier map of the difference function $F_x - F$ in the plane of

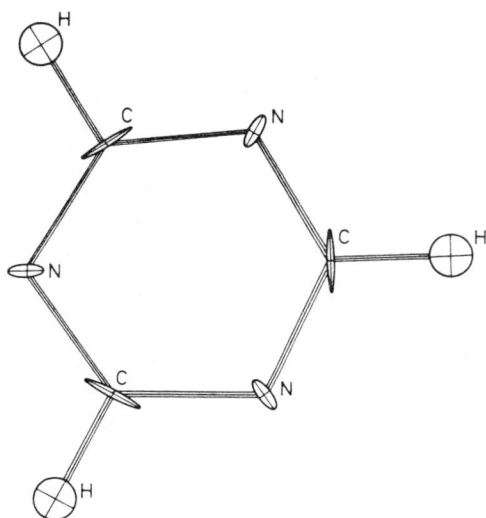

Figure 6.4 A plot of the difference between the thermal ellipsoids determined by X-rays and neutrons for the carbon and nitrogen atoms in sym-triazine. The hydrogen atoms in the molecules are indicated as circles (From Coppens[8], by courtesy of Acta Cryst.)

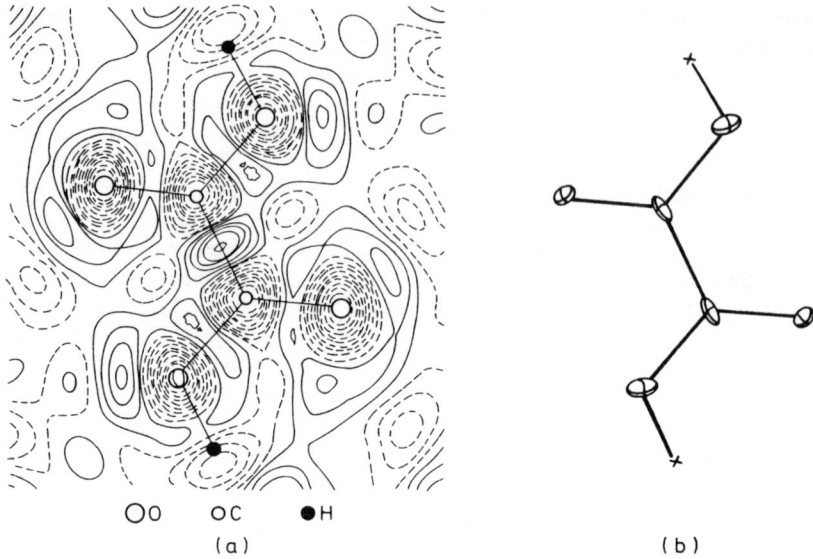

Figure 6.5 (a) shows an X–N map through the molecular plane of α-deuterated oxalic acid dihydrate, indicating concentrations of electron density at the centres of the C–C and C–O bonds and lone pairs near the carbonyl oxygen atoms and the oxygen atoms of the hydroxyl groups. (b) shows the difference between the thermal ellipsoids of the non-hydrogen atoms as assessed by X-rays and neutrons (From Coppens et al.[9], by courtesy of Acta Cryst.)

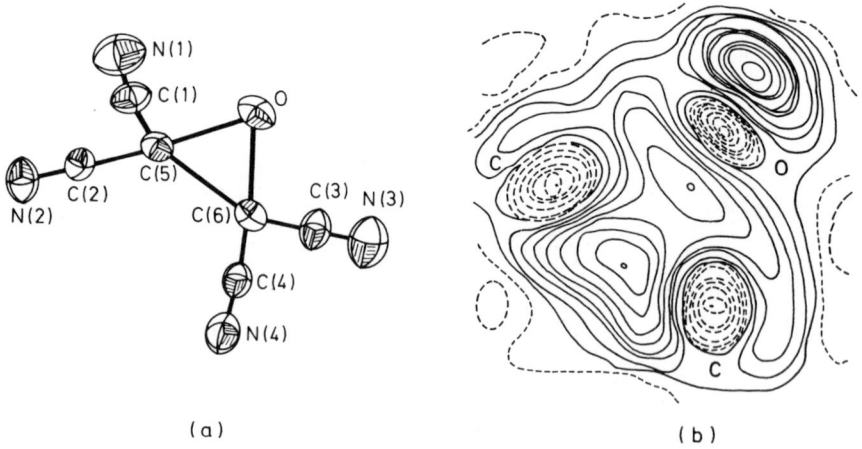

Figure 6.6 (a) The molecular structure and thermal ellipsoids for tetracyanoethylene oxide determined with neutrons, and (b) an X–N synthesis of the electron density in the plane of the ethylene oxide ring (From Matthews et al.[11] and Matthews and Stucky[10], by courtesy of J. Amer. Chem. Soc.)

the ethylene oxide ring. The map is notable for the depressions of electron density at the atomic centres and for the excess of density at the bond centres and the lone pair of the oxygen atom. *Table 6.1* from the paper of Matthews and Stucky[10] shows how the thermal parameters measured from X-ray data are falsely increased in order to simulate the bonding and lone-pair density which is not accounted for in the spherical model. It will be seen that the diagonal elements of the anisotropic thermal tensors have X-ray values which are greater than the neutron values by amounts which range from 2 to 9 times their standard deviations.

Apart from the direct demonstration of residual electron density in these so-called 'X minus N' maps, the effect of asphericity is indicated

Table 6.1 Diagonal thermal parameters* of tetracyanoethylene for X-rays and neutrons, $b_{ij} \times 10^4$

Atom	ij	X-ray	Neutron	Δ	Δ/σ
Oxygen	11	110 (1)	94 (4)	16 (4)	4.0
	22	242 (3)	198 (12)	44 (12)	3.7
	33	75 (1)	66 (3)	9 (3)	3.0
Carbon (1)	11	118 (2)	109 (4)	9 (4)	2.3
	22	196 (3)	168 (9)	28 (10)	2.8
	33	72 (1)	54 (2)	18 (2)	9.0
Nitrogen (3)	11	155 (2)	121 (3)	34 (4)	8.5
	22	354 (4)	294 (9)	60 (10)	6.0
	33	98 (1)	89 (2)	9 (2)	4.5
Carbon (6)	11	100 (2)	88 (3)	12 (4)	3.0
	22	198 (3)	171 (8)	26 (9)	2.9
	33	55 (1)	40 (2)	15 (2)	7.5

*The temperature factor is expressed as
$\exp-(b_{11}h^2 + b_{22}k^2 + b_{33}l^2 + 2\beta_{12}hk + 2\beta_{13}hl + 2\beta_{23}kl)$

Table 6.2 Shift of X-ray positions relative to nuclear positions given by neutrons, for carbon, nitrogen and oxygen atoms

Compound	Atom	Shift/Å	Direction	Reference
sym-Triazine	C in ring	0.015(8)	Away from C–H bond	Coppens[7]
	N in ring	0.009(6)	Towards lone pair	
Cyanuric acid	O(1) in C=O	0.006(1)	Towards lone pairs	Coppens and Vos[5]
	O(2) in C=O	0.003(1)	Towards lone pairs	
Hexamethylene-tetramine	N	0.021(7)	Towards lone pair	Duckworth et al.[1]
a-Deuterated oxalic acid dihydrate	O(1) in C–O–D	0.008(2)	Away from O–D	Coppens et al.[9]
a-Oxalic acid dihydrate	O(1) in C–O–H	0.008(2)	Away from O–H bond	Coppens et al.[9]

Table 6.3 Comparison of bond lengths measured with X-rays and neutrons, for X–H bonds, in Å

Compound	Bond	Neutron	X-ray	Difference	Reference
a-Oxalic acid dihydrate	O(1)–H(1)	1.026(7)	0.89(2)	0.14(2)	Coppens et al.[9]
	O(3)–H(2)	0.964(7)	0.84(2)	0.12(2)	
	O(3)–H(3)	0.956(9)	0.79(3)	0.17(3)	
a-Deuterated oxalic acid dihydrate	O(1)–D(1)	1.031(2)	0.86(2)	0.17(2)	Coppens et al.[9]
	O(3)–D(2)	0.954(2)	0.83(2)	0.12(2)	
	O(3)–D(3)	0.954(2)	0.78(2)	0.17(3)	
sym-Triazine	C–H	1.045(16)	0.92(4)	0.13(4)	Coppens[7]
Sucrose	C–H	1.095	0.96	0.13(1)	Brown and Levy[12]
	O–H	0.968	0.79	0.18(3)	
Cyanuric acid	N(1)–H(1)	1.026(2)	0.79(3)	0.24(3)	Coppens and Vos[5]
	N(2)–H(2)	1.037(1)	0.87(1)	0.17(2)	

in a comparison of the atomic positions and bond lengths obtained separately in the neutron and X-ray refinements. Some examples of these differences are given in *Tables 6.2* and *6.3*. For hydrogen, differences of up to 0.2 Å have been found, but they are much smaller for heavier atoms and even for C, N and O the differences are no greater than about 0.1 Å.

At the present time the accuracy of determination of electron density by the X-N method seems comparable to that of theoretical computation for relatively simple substances. A direct comparison of theory and experiment has been provided by Thomas, Tellgren and Almlöf[13] for lithium formate monohydrate, $LiHCOO \cdot H_2O$. In this substance the water molecule takes part in two hydrogen bonds of significantly different strength; the O–H distances are 1.74 and 1.95 Å respectively and the corresponding O–O distances are 2.71 and 2.90 Å. Both experiment and calculation show a marked difference of electron density for these two bonds. In particular there is a marked overall loss of charge in the whole region around the hydrogen atom which takes part in the stronger bond. This paper includes comments about the sources of error in X-N maps, particularly in relation to non-centrosymmetric structures. A discussion of the experimental accuracies has also recently been given by Coppens[14] in a review of the substances which have been examined so far.

References

1. DUCKWORTH, J.A.K., WILLIS, B.T.M. and PAWLEY, G.S., *Acta Cryst.,* **A25,** 482 (1969)
2. DUCKWORTH, J.A.K., WILLIS, B.T.M. and PAWLEY, G.S., *Acta Cryst.,* **A26,** 263 (1970)
3. BECKA, L.N. and CRUICKSHANK, D.W.J., *Proc. Roy. Soc.,* **A173,** 435 (1963)

4. COPPENS, P. and COULSON, C.A., *Acta Cryst.,* **23,** 718 (1967)
5. COPPENS, P. and VOS, A., *Acta Cryst.,* **B27,** 146 (1971)
6. VERSCHOOR, G.C. and KEULEN, E., *Acta Cryst.,* **B27,** 134 (1971)
7. COPPENS, P., *Science,* **158,** 1577 (1967)
8. COPPENS, P., *Acta Cryst.,* **B24,** 1272 (1968)
9. COPPENS, P., SABINE, T.M., DELAPLANE, R.G. and IBERS, J.A., *Acta Cryst.,* **B25,** 2451 (1969)
10. MATTHEWS, D.A. and STUCKY, G.D., *J. Amer. Chem. Soc.,* **93,** 5954 (1971)
11. MATTHEWS, D.A., SWANSON, J., MUELLER, M.H. and STUCKY, G.D., *J. Amer. Chem. Soc.,* **93,** 5945 (1971)
12. BROWN, G.M. and LEVY, H.A., *Acta Cryst.,* **B29,** 790 (1973)
13. THOMAS, J.O., TELLGREN, R. and ALMLÖF, J., *Acta Cryst.,* **B31,** 1946 (1975)
14. COPPENS, P., *Acta Cryst.,* **B30,** 255 (1974)

7

STUDIES OF BIOLOGICAL MATERIALS

The first biological material for which a structural analysis by neutrons was reported was the monocarboxylic acid derivative of vitamin B_{12}, $C_{63}H_{87}O_{15}N_{13}PCo \cdot 16H_2O$, which was studied by Moore, Willis and Hodgkin[1], with further unpublished work by Hodgkin et al. This material has a monoclinic unit cell measuring 14.9 × 17.5 × 16.4 Å, with $\beta = 104°$, and a neutron wavelength of 1.53 Å was used, rather longer than usual in order to improve the angular resolution of the closely-spaced reflections. There are 228 atoms in the asymmetric unit and about 3000 reflections were measured, going down to an interplanar spacing of 1 Å. Using Fourier and least-squares methods an R value of 18% was achieved. The detail achieved in the Fourier projections is indicated by the portions shown in *Figure 7.1*. As expected, the hydrogen atoms show up prominently. Two other points are noteworthy. First, diagram (a) indicates how the cobalt atom is a light atom for neutrons, whereas for X-rays it is by far the most prominent atom in the structure. Secondly, the ability of neutrons to distinguish nitrogen readily from oxygen, because of its significantly larger scattering, together with the detail of the hydrogen atoms, means that the group of atoms in (b) is conclusively identified as $-CH_2CONH_2$ rather than $-CH_2COOH$, a distinction which is not easy to make with X-rays.

At a very ambitious level Schoenborn[2] has carried out a feasibility study of the possibility of a full structural analysis of a protein, using myoglobin in which most of the water of crystallization had been deuterated in order to reduce the incoherent background-scattering. With

Figure 7.1 Portions of the neutron scattering density of a monocarboxylic acid derivative of vitamin B_{12}. (a) emphasizes a cobalt atom at the centre of the corrin nucleus and (b) conclusively identifies a group of atoms as $-CH_2CONH_2$, rather than $-CH_2COOH$ (From Moore, Willis and Hodgkin[1], by courtesy of Nature)

a unit cell which measures 64.5 × 30.9 × 34.8 Å, with $\beta = 106°$, and a molecular weight of about 18 000, the reflections are very closely spaced and 25 000 need to be measured to provide spatial resolution of 1.5 Å. In the preliminary experiment 3200 independent reflections covered spacings down to 2.8 Å and the resulting projections, which incorporated the X-ray phases, revealed the main features of the molecules. The side groups were defined better than with X-rays because of the details of the hydrogen atoms. It was concluded that a full analysis would be practicable with a high-flux reactor, though necessarily with very long counting times for weak reflections.

At the present time a variety of exploratory experiments are being carried out, ranging over diverse topics such as lipid bilayers, fibrous proteins such as muscle and collagen and dynamical studies of large molecules using inelastic scattering. At the same time the potential of two quite separate applications in the biological field has been firmly established. The first of these exploits the anomalous scattering of neutrons, which we mentioned in Chapter 2, as a method of solving the phase problem in structural analysis. The second uses the small-angle scattering of neutrons to determine the sizes and shapes of large biological molecules in solution.

7.1 Anomalous scattering

We noted in Chapter 2 that in the neighbourhood of a scattering resonance the scattering amplitude becomes a complex quantity, with a phase change which is not simply 0° or 180° but takes some intermediate value. For most elements and isotopes the resonances are so far away from thermal energies that the anomalous scattering is unimportant in diffraction studies, but exceptions such as ^{113}Cd and ^{149}Sm do exist. In these cases very marked effects occur in the neighbourhood of a wavelength of 1 Å and they are immensely more substantial than the anomalous scattering effects which take place for X-rays in the neighbourhood of an absorption edge. *Figure 7.2* shows the calculated values of the real and imaginary components of the scattering amplitude of ^{113}Cd: we emphasize in particular that for $\lambda = 0.7$ Å the imaginary contribution b'' is seven times as large as the scattering amplitude at short wavelengths, which has a real value of 0.70×10^{-12} cm. The consequence of this large value of b'' is that in non-centrosymmetric structures the intensities of the reflections hkl, $\overline{hkl}$, i.e. the reflections from opposite sides of a crystallographic plane, will no longer be of equal intensity. It can be shown (Dale and Willis[3]) that measurement of the two different intensities leads to a knowledge of the phase angle of the reflection, assuming that an ambiguity between two possible values $|F|, a_1$ and $|F|, a_2$ is resolved by making measurements at two different wavelengths, one on each side of the resonance position.

Table 7.1 lists the isotopes which are applicable to this method, together with their resonance wavelengths and their abundances in the natural elements which contain them. The structure of cadmium nitrate

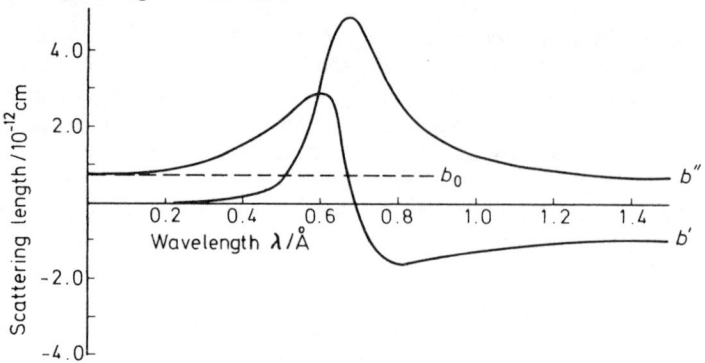

Figure 7.2 *Calculated values of the real and imaginary components b' and b'' of the scattering amplitude of ^{113}Cd as a function of neutron wavelength, for the parameters measured by Brockhouse*

tetradeuterate has been determined by Macdonald and Sikka[4]. In this case the neutron intensity at wavelengths shorter than the ^{113}Cd resonance at 0.68 Å was insufficient to permit the use of the multiple-wavelength method of resolving the phase ambiguity and the analysis was achieved by using the double-phased Fourier synthesis method of Ramaseshan[5] and the sine-Patterson procedure of Okaya and Pepinsky[6]. It would be possible to use the multiple-wavelength method if the intensity of short-wavelength neutrons were increased by the use of a 'hot source' in the reactor, as at the Institut Laue-Langevin at Grenoble. It will be noted from *Table 7.1* that the resonant wavelength for ^{149}Sm is substantially greater, at 0.92 Å, than for ^{113}Cd and measurements are being made by Hodgkin and her co-workers to improve the knowledge of the structure of insulin (Adams *et al.*[7]) by incorporating the ^{149}Sm isotope in the insulin molecule.

7.2 Small-angle scattering

The study of the sizes and shapes of biological molecules rests on the basic assessment of small-angle scattering by ordinary optical diffraction theory. It can be shown[8] that the scattering by a collection of particles will fall off with the scattering angle 2θ according to the expression

$$I = I_0 \exp(-Q^2 R^2/3) \qquad (7.1)$$

Q is the momentum transfer vector, with a numerical value equal to $(4\pi \sin \theta)/\lambda$, which approximates to $4\pi\theta/\lambda$ at small angles, and R is the radius of gyration for scattering. By analogy with classical dynamics, R^2 is defined as the mean value of the square of the distance of each scattering atom from an axis drawn through the centre of gravity and parallel to the scattering vector, assuming that each contribution to the summation

Table 7.1 Data for anomalously scattering isotopes

Element or isotope	Natural abundance	Resonant wavelength/Å	$b''/10^{-12}$ cm	
			$\lambda = 1$ Å	At resonance
Cd		0.68	0.15	0.58
^{113}Cd	0.12	0.68	1.2	4.70
Sm		0.92	0.7	0.88
^{149}Sm	0.14	0.92	5.1	6.30
Eu		0.6	0.07	1.31
^{151}Eu	0.48	0.6	0.15	2.74
Gd		1.8	0.8	1.2
^{157}Gd	0.16	1.8	4.0	6.60

is weighted by the atomic scattering amplitude of the atom. It follows from equation 7.1 that

$$\log I = \text{constant} - \frac{Q^2 R^2}{3}$$

$$= \text{constant} - \frac{16\pi^2 \theta^2}{3\lambda^2} R^2 \qquad (7.2)$$

so that a plot of $\log I$ against θ^2 will give a value for R.

In the case of the scattering by macromolecules in solution the scattering depends on the amount by which the resultant scattering amplitude of the molecule exceeds the scattering of the molecule which it displaces. Accordingly the scattering per unit solid angle, at a particular value of Q or θ, is given by

$$\left(\frac{d\sigma}{d\Omega}\right)_Q = \left| \sum_i b_i \exp(i\mathbf{Q}\cdot\mathbf{r}_i) - \rho_s \int_V \exp(i\mathbf{Q}\cdot\mathbf{r})d\mathbf{r} \right|^2 \qquad (7.3)$$

where ρ_s is the scattering-length density of the solvent. The summation is made over the various atoms in the molecule, of scattering length b_i and position vector $\mathbf{r}_i$, and the integral is taken over the volume of solvent excluded by the molecule. We shall find that the advantages of studying the small-angle scattering will arise because in aqueous solutions we can control the value of ρ_s by varying the hydrogen/deuterium ratio in the water.

It is convenient to define $\rho(\mathbf{r})$ as the scattering-length density at any position $\mathbf{r}$ in the molecule and to re-write equation 7.3 in the form

$$\left(\frac{d\sigma}{d\Omega}\right)_Q = \left| \int_V \rho(\mathbf{r}) \exp(i\mathbf{Q}\cdot\mathbf{r})d\mathbf{r} - \rho_s \int_V \exp(i\mathbf{Q}\cdot\mathbf{r})d\mathbf{r} \right|^2 \qquad (7.4)$$

Stuhrmann[9] has shown that it is advantageous to divide the scattering into two new components dependent respectively on the external shape of the molecule and on its internal fluctuations of scattering-length density. A quantity ρ_M is defined as the *average* value of the scattering-length density over the molecule and equation 7.4 is re-written in the equivalent form

$$\left(\frac{d\sigma}{d\Omega}\right)_Q = \left| \int_V (\rho(r) - \rho_M) \exp(iQ\cdot r)dr + (\rho_M - \rho_s) \exp(iQ\cdot r)dr \right|^2 \quad (7.5)$$

$$= (\rho_M V)^2 \left| \frac{1}{\rho_M V} \int_V (\rho(r) - \rho_M) \exp(iQ\cdot r)dr + \frac{\rho_M - \rho_s}{\rho_M} \frac{1}{V} \int \exp(iQ\cdot r)dr \right|^2 \quad (7.6)$$

This can then be written in terms of two form-factors, $F_{Mol}(Q)$ and $F_{Shape}(Q)$, defined by

$$F_{Mol}(Q) = \frac{1}{\rho_M V} \int_V (\rho(r) - \rho_M) \exp(iQ\cdot r)dr \quad (7.7)$$

and

$$F_{Shape}(Q) = \frac{1}{V} \int_V \exp(iQ\cdot r)dr \quad (7.8)$$

as

$$\left(\frac{d\sigma}{d\Omega}\right)_Q = (\rho_M V)^2 \left(F_{Mol}(Q) + \frac{\rho_M - \rho_s}{\rho_M} F_{Shape}(Q)\right)^2 \quad (7.9)$$

The significance of F_{Mol} and F_{Shape}, as defined in these equations, can be appreciated with the aid of *Figure 7.3*. F_{Mol} depends on $\rho(r) - \rho_M$, which is equivalent to $\rho_f(r)$ in the figure and represents the fluctuation of the scattering-length density within the molecule; the scattering level AA' which determines the values of $\bar{\rho}$ and ρ_M is chosen such that

$$\int_V \rho_f(r)dr = 0$$

F_{Shape} depends solely on the *shape* of the molecule and its multiplying factor $\rho_M - \rho_s$, in equation 7.9, is the quantity $\bar{\rho}$ in the figure. $\bar{\rho}$ will depend on the hydrogen/deuterium ratio in the solvent and is termed the 'contrast', being the difference between the mean-scattering-length density of the molecule and the solvent.

By making measurements in solvents containing different ratios of hydrogen and deuterium it is possible to determine the values of F_{Mol} and F_{Shape}. It follows from equation 7.3 that the value of the scattering when extrapolated to $Q = 0$ will be proportional to

$$\left[\sum b_i - \rho_s V\right]^2$$

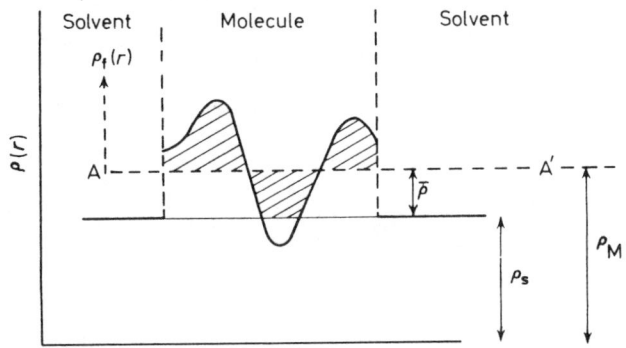

Figure 7.3 Representation of the scattering lengths for a molecule surrounded by solvent. The ordinate $\rho(r)$ gives the absolute scattering-length density over the molecule and ρ_S as its value in the solvent: ρ_M is the average value of $\rho(r)$ over the molecule and the ordinate $\rho_f(r)$ denotes the excess of $\rho(r)$, at any point, above the average

and the square root of this intensity will be a linear function of ρ_S which passes through zero when

$$\sum b_i = \rho_S V$$

i.e. when $\rho_S = \rho_M$. In this way the value of ρ_M can be determined and Figure 7.4 shows some data for haemoglobin[9]: the zero intensity occurs for a heavy-water concentration of 40%, for which ρ_S, and hence ρ_M, is equal to 2.24×10^{10} cm^{-2}. The figure also shows some results for ferritin for which the intensity ordinate does *not* fall to zero; this is accounted for by the variable iron content of the ferritin molecules, leading to a broad distribution function, and not a single defined value, for the scattering-length density. It can be seen from equation 7.9 that when $(\rho_M - \rho_S)/\rho_M$ is large, i.e. for high contrast, then $(d\sigma/d\Omega)_Q$ will be determined solely by F_{Shape}. Accordingly a value of R^2 determined, from equation 7.2, under these conditions will be representative of the molecular shape alone. An example of such a determination[9] is illustrated in Figure 7.5, where the measured value of R^2 is plotted against the reciprocal of $\rho_S - \rho_M$. For zero value of the abscissa, R^2 equals about 200 Å^2 and this value will represent the molecular shape. In a similar way we see from equation 7.9 that the value of R^2 when $\rho_M - \rho_S$ is zero will correspond to F_{Mol}, the form-factor for the fluctuations above the average of the scattering-length density within the molecule.

The measurements of small-angle scattering which we have just described have only become possible through the development of special neutron instruments which permit observations at very small values of 2θ, down to a few minutes of arc, and ranging over values of Q from 0.001 to 1 Å^{-1}. The most powerful and versatile instrument[11] exists at the Institut Laue-Langevin at Grenoble and utilizes curved guide-tubes which not only ensure very highly collimated beams of high intensity but also

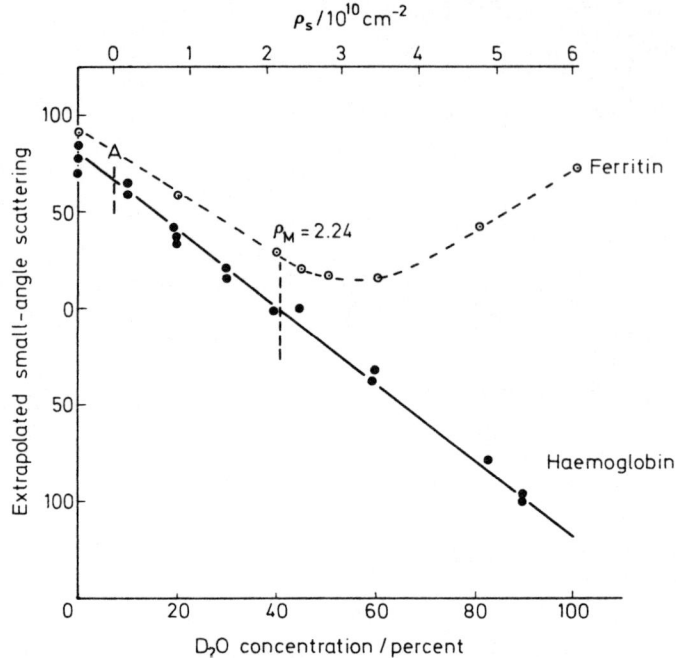

Figure 7.4 The variation with D_2O content in heavy/light water mixtures, of the small-angle scattering of haemoglobin and ferratin extrapolated to $Q = 0$. The scale at the top of the diagram gives the scattering-length density ρ_s of the solvent. For haemoglobin the scattering is zero when ρ_s equals 2.24×10^{10} cm^{-2}; for ferratin the scattering never falls to zero (After Schelten et al.[10], by courtesy of J. Biol. Chem.)

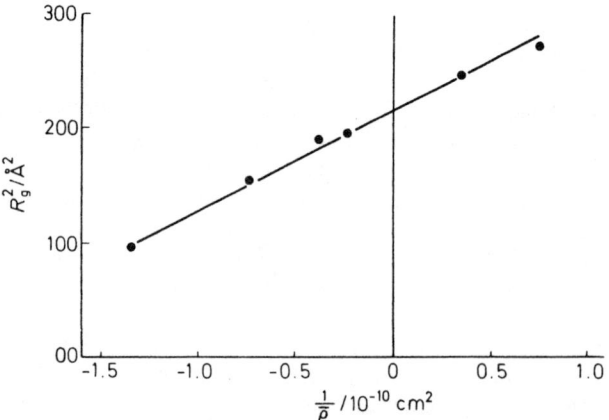

Figure 7.5 Measurements of the square of the radius of gyration for scattering by myoglobin in heavy/light water mixtures. The abscissa is the reciprocal of $\bar{\rho}$, which is the difference between ρ_M and ρ_S (From the results of Stuhrmann[9])

produce freedom from fast-neutrons and γ-rays. A range of incident wavelengths from 2 to 20 Å is available, as chosen by a mechanical velocity selector, and distances of up to 40 m can be employed between monochromator and sample and between sample and detector, to give the maximum angular resolution. The same instrument is being used for obtaining structural information from semi-crystalline materials such as the fibrous proteins muscle and collagen which have very large repeating units, in the region up to 700 Å, though imperfectly arranged. A quite simple diffractometer for this kind of work has also been described and used by Haywood and Worcester[12]. *Figure 7.6* indicates how measurements in the region of 2θ from zero to 6° reveal four orders of diffraction from a sample of dog sciatic nerve. This instrument provided an incident beam with an overall width of 0.8° at a wavelength of 4.7 Å, from a beryllium filter, and the angular positions in *Figure 7.6* correspond to a repeat distance of 180 Å. A study with this instrument of the diffraction from hydrated multilayers of a mixture of egg lecithin and cholesterol, oriented on glass substrates, has been reported by Worcester and Franks[13]. At Grenoble the observations are made with a two-dimensional position-sensitive detector effectively consisting of a grid of 64 × 64 BF_3 counters, each with a sensitive area of about 1 cm². Accordingly a two-dimensional plot of the diffracted beam is obtained and *Figure 7.7* shows an example. This is a pattern for frog sartorious muscle obtained by Worcester *et al.*[14] using a wavelength of 7.0 Å. The muscle fibres are mounted vertically and the strong reflection at the top and bottom is the meridional reflection of 143 Å Bragg spacing (A). Meridional reflections are also observed at about 216 Å (B) and 440 Å (C). Off-meridional intensity is distinct on the first (D) and third (E) layer-lines. These features of the diffraction pattern arise mainly from the helical arrangement of myosin cross bridges protruding from the

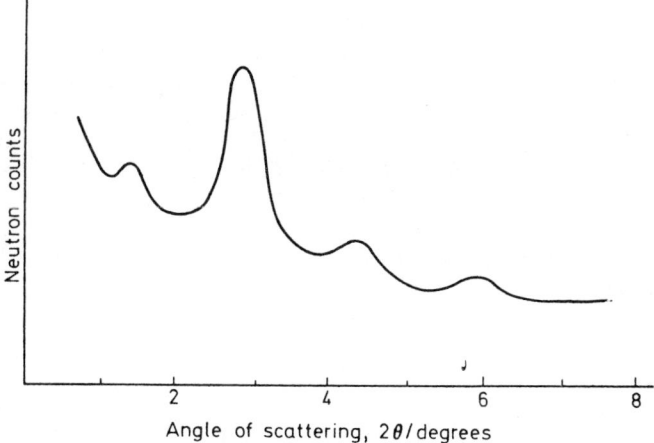

Figure 7.6 The scattering by dog sciatic nerve over the angular range from 0-8° for neutrons with λ = 4.7 Å (From Haywood and Worcester[12]*, by courtesy of J. Phys. E)*

90 Studies of Biological Materials

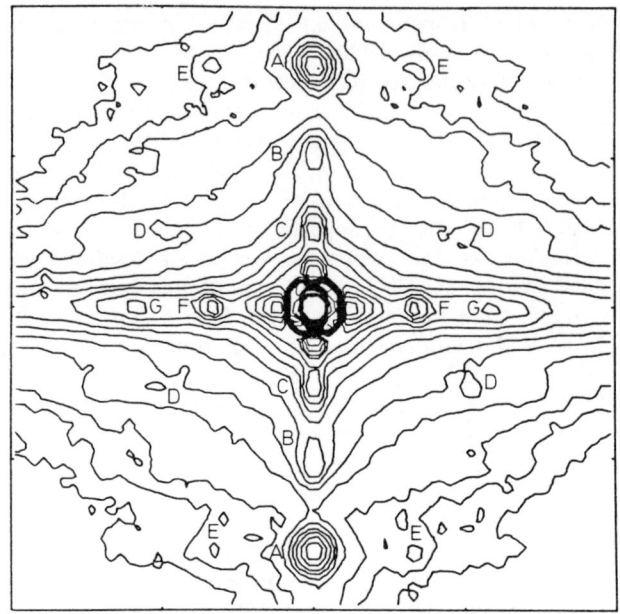

Figure 7.7 Low-angle neutron diffraction from a living frog sartorius muscle in D_2O Ringer solution is shown in this figure as a contour map of data obtained with the 64 × 64 cm² area detector at the Institut Laue-Langevin, Grenoble. The distance from the sample to the 64 × 64 cm² detector is 540.5 cm; the wavelength, 7.0 Å (±4%); and the exposure time, 1 hour. The contours of the number of counts in each 1 cm² detector element increase by a factor of $\sqrt{2}$ from the lowest contour (512 counts cm⁻²) with the exception of the second contour (615 counts cm⁻²) (Figure provided by courtesy of D.L. Worcester, Atomic Energy Research Establishment, Harwell; J.M. Gillis, University of Louvain; E.J. O'Brien, Medical Research Council Cell Biophysics Unit, London; and K. Ibel, Institut Laue-Langevin, Grenoble)

myosin thick filaments. The equatorial reflections (F and G) can be indexed as the 100 and 110 reflections of the hexagonal lattice of myosin filaments. The four very low-angle contour peaks (unlettered) are not true reflections, but arise from the increasing intensity of the central scatter, being cut off on the four sides of the beam stop. These measurements were made in aqueous solutions using various mixtures of light and heavy water. The quality of the pattern depends considerably on the hydrogen/deuterium ratio, indicating that it is the scattering contrast between water and protein that is important.

References

1. MOORE, F.H., WILLIS, B.T.M. and HODGKIN, D.C., *Nature*, **214,** 130 (1967)
2. SCHOENBORN, B.P., *Nature,* **224,** 143 (1969)
3. DALE, D.H. and WILLIS, B.T.M., *AERE Report R5195* (1966)

4. MACDONALD, A.C. and SIKKA, S.K., *Acta Cryst.*, **B25**, 1804 (1969)
5. RAMASESHAN, S., in *Advanced Methods of Crystallography*, ed. G.N. Ramachandran, 67, Academic Press, London (1964)
6. OKAYA, Y. and PEPINSKY, R., in *Computing Methods and the Phase Problem in X-ray Crystal Analysis,* ed. R. Pepinsky, J.M. Robertson and J.C. Speakman, Pergamon Press, Oxford (1961)
7. ADAMS, M.J., BLUNDELL, T.L., DODSON, E.J., DODSON, G.G., VIJAYAN, M., BAKER, E.N., HARDING, M.M., HODGKIN, D.C., RIMMER, B. and SHEAT, S., *Nature*, **224**, 491 (1969)
8. GUINIER, A., *X-ray Diffraction,* Freeman & Co., London (1963)
9. STUHRMANN, H.B., *J. Appl. Cryst.,* **7**, 173 (1974)
10. SCHELTEN, J., SCHLECHT, P., SCHMATZ, W. and MAYER, A., *J. Biol. Chem.,* **247**, 5436 (1972)
11. SCHMATZ, W., SPRINGER, T., SCHELTEN, J. and IBEL, K., *J. Appl. Cryst.,* **7**, 96 (1974)
12. HAYWOOD, B.C.G. and WORCESTER, D.L., *J. Phys. E,* **6**, 568 (1973)
13. WORCESTER, D.L. and FRANKS, N.P., *J. Mol. Biol.,* **100**, 359 (1976)
14. WORCESTER, D.L., GILLIS, J.M., O'BRIEN, E.J. and IBEL, K., *Neutrons in Biology: Proceedings of 1975 Brookhaven Biology Symposium*

8

MEASUREMENTS OF COVALENCY

In a fully ionic compound such as sodium chloride the ionic bond is achieved by complete transfer of the one outer electron of the sodium atom to complete the valency shell of the chlorine atom. In other compounds, particularly the salts and oxides of the first transition series, such as NiO, the electron transfer is not complete and the ionic bond may be considered to be contaminated by a small amount of covalency. In the case of magnetic materials, in which the cations bear electrons of unpaired spin providing magnetic moment, any partial transfer of electrons back to the cation will both affect the value of the latter's magnetic moment and will produce some unpaired spin on the anion ligand. These two features will necessarily alter the magnetic scattering of neutrons by the material and accordingly it should be possible to detect and measure the degree of covalency from observations of the magnetic scattering[1]. *Figure 8.1* illustrates the situation for a hypothetical isolated pair of cation and anion and indicates in particular how the fraction

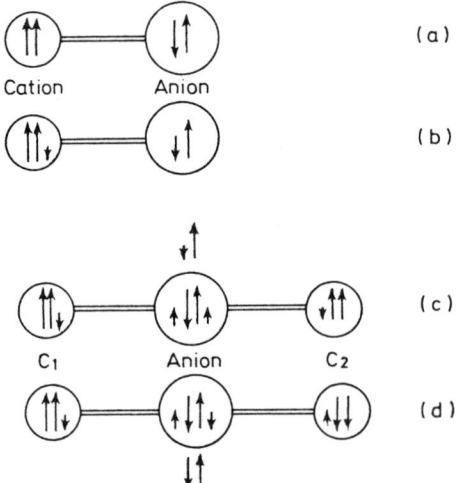

Figure 8.1 (a) illustrates a purely ionic bond between cation and anion in contrast to (b), where a proportion of downward pointing spin is transferred, by covalency, from anion to cation. In (c) the anion effectively receives spin of the same sign from its two neighbouring cations C_1 and C_2 which are in ferromagnetic alignment; in (d) no net spin appears on the anion because the two cations are in antiferromagnetic alignment

of electron returned to the cation will have a spin of opposite sign to that existing initially on the cation. It will be evident that both the magnitude and angular distribution of magnetic scattering for the pair in (b) will be different from what would occur for (a). We shall in fact show that there will be changes in magnetic form-factors and in certain cases magnetic scattering will be found at Bragg angles which would otherwise be prohibited.

In an actual magnetic material each cation will be surrounded by several anions, each of which will indulge in charge transfer, so that the cation spin will be proportionately reduced. However, the effect on the anion will not be cumulative in the same way, since the cations which are neighbours to a given anion may have their spins oriented differently, depending on the magnetic structure of the material. Thus in *Figure 8.1(c)*, where the two neighbours C_1 and C_2 of the anion are aligned ferromagnetically, twice as large a component of spin will appear on the anion as that which occurred in (b). On the other hand in (d), where the two neighbours of the anion are aligned antiferromagnetically, the effects contributed by C_1 and C_2 are equal and opposite and no spin appears on the anion. However, in less symmetrical antiferromagnetic structures, and MnF_2 is an example, it is possible for an anion to have two cation neighbours of +ve sign and a third neighbour of −ve sign: in such a case a finite spin would indeed appear on the anion and affect the neutron scattering.

We shall proceed to consider particular examples of magnetic materials, to correlate the scattering with the magnetic structure and to show how the degree of covalency of the bond can be computed from the observations. In order to permit this, however, we will first express quantitatively the magnetic moment density, in the neighbourhood of a cation, in terms of the covalency parameters. For the simple case of a cation which is at a centre of symmetry between two nearest-neighbour anions the electron spin density can be expressed by writing the wave function as a linear combination of a d function for the metal ion and a p function for each ligand (see Marshall and Lovesey[2]) as

$$D(r) = d^2(r)[1 + 4A_\sigma S_\sigma - 2A_\sigma^2]$$
$$- 2A_\sigma d(r)[p(r) - p'(r)] \qquad (8.1)$$
$$+ A_\sigma^2 [p^2(r) + p'^2(r)]$$

where $d(r)$, $p(r)$, $p'(r)$ denote the d orbital of the magnetic ion and the p orbitals of the two neighbouring ligands. S_σ is an overlap integral between the d and p orbitals and A_σ is a covalency parameter. For a general account of the molecular orbital description of the bonding between cation and anion the reader is referred to a paper by Owen and Thornley[3] and an extensive paper by Tofield[4], which also includes a detailed account of the study of covalency by neutron scattering and a review of recent progress. The expression in equation 8.1 divides naturally into three parts: the first term is confined solely to the parent magnetic ion, the second concerns the overlap density between the ions and the third term gives spin density which is entirely on the ligands and which, as we have seen above, vanishes in symmetrical situations. The above expression can be re-written as

$$D(r) = d^2(r)[1 - 2A_\sigma^2] + 2A_\sigma[2S_\sigma d^2(r) - d(r)p(r) + d(r)p'(r)]$$
$$+ A_\sigma^2[p^2(r) + p'^2(r)] \tag{8.2}$$

in contrast to $D(r) = d^2(r)$ for a free cation. The first term now corresponds to a moment of $1 - 2A_\sigma^2$, in contrast to unity for a free cation, and the second term is positive near the centre of the cation, negative near its boundaries but overall has a zero integrated value. The form-factor for the magnetic scattering will be derived by summing the Fourier transforms of these three terms. The result is summarized in *Figure 8.2*.

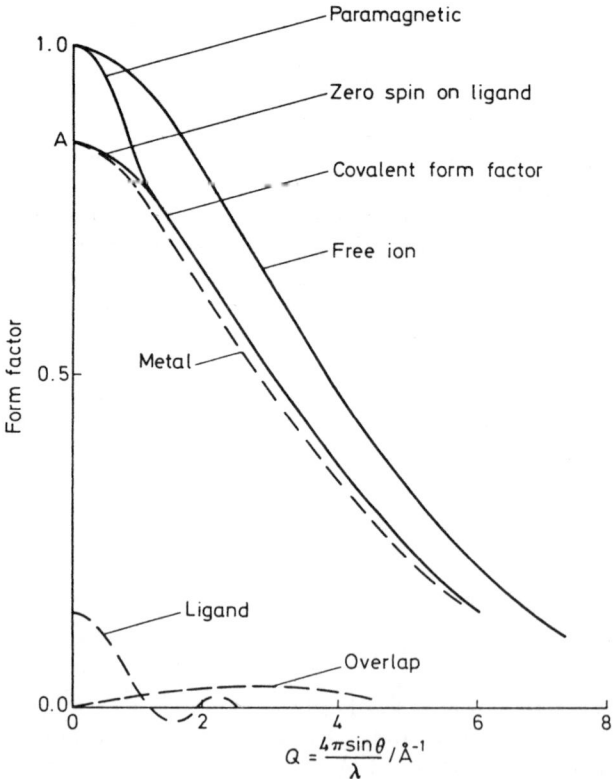

Figure 8.2 The effect of covalency on the form-factor of a magnetic cation. Transfer of oppositely directed spin to the cation results in a reduction of its moments (indicated by the fall of point A below unity) and a slight broadening of the form-factor. For an isolated paramagnetic ion, or in a ferromagnetic material, the net charge on the ligand adds a forward-going peak to the curve which restores it to unity at $Q = 0$; in most antiferromagnetic materials, e.g. NiO, there is no net ligand charge and this 'ligand peak' is ineffective (From Jacobson[20], by courtesy of Oxford University Press)

The main part of the form-factor, contributed by the first term, is of the same shape as for a free ion, but reduced in magnitude by the factor $1 - 2A_\sigma^2$. The second term gives an overlap contribution to the form-factor which is zero at $Q = 0$, because of the net overall moment involved, but then rises to a maximum before decreasing again. When these two terms are added together the resultant form-factor is slightly flatter than that for the free ion and accordingly has a slightly larger half-width. The ligand contribution to the form-factor, for this particular case of a cation between a pair of anions, is also shown in *Figure 8.2*. Whether such a contribution occurs in a practical magnetic material will depend, as we have already shown, on the environment of the anion.

In three dimensions a magnetic ion is typically surrounded by an approximately regular octahedron of six ligand ions and the corresponding calculation involves three parameters A_σ, A_s and A_π which describe the transfer of charge into the appropriate orbitals. The factor which reduces the free ion form-factor in *Figure 8.2* can be shown to be $1 - 3A_\sigma^2 - 3A_s^2$ for e_g type orbitals and $1 - 4A_\pi^2$ for t_{2g} orbitals. As an example we may consider the Ni^{2+} ion in $KNiF_3$ and NiO, which is associated with e_g orbitals. Accordingly the neutron data should give a value for $A_\sigma^2 + A_s^2$ and this could then be compared with nuclear magnetic resonance data, which, as we shall see later, yield the same quantity. Hutchings and Guggenheim[5] examined $KNiF_3$ with neutrons and derived a value of 2.6 ± 1.8% for $A_\sigma^2 + A_s^2$, in comparison with a more accurate resonance value of 4.3 ± 0.25%. NiO has been much studied by neutrons and it was the original measurements of it by Alperin[6] which led to the realization that neutron intensities and form-factors were affected by covalency in the manner which we have been discussing. Alperin made absolute measurements of intensity, by comparison with nuclear intensities, for a single crystal and his results are indicated in *Figure 8.3* as a form-factor curve. Extrapolation to $Q = 0$ gave a value of 1.81 ± 0.20 μ_B for the magnetic moment of Ni^{2+}, in comparison with an expected value of 2.23 μ_B (allowing for some orbital contribution) for the free ion. We emphasize that smooth extrapolation of the form-factor curve to $Q = 0$ is reliable in this case, since the magnetic symmetry is such that the oxygen atoms are surrounded by equal numbers of parallel and antiparallel spins on the Ni^{2+} ions and there is no net contribution to the form-factor from the ligands. The values just given mean that covalency has reduced the magnetic moment by a factor of 0.82 ± 0.10, so that the value of $3A_\sigma^2 + 3A_s^2$ is 0.18 ± 0.10. Correction for other effects reduced this to 0.12 ± 0.10, meaning that the magnitude of $A_\sigma^2 + A_s^2$ is 4%. Very similar results have been obtained more recently by Fender, Jacobsen and Wedgwood[7]. It is to be emphasized that this effect of covalency is by no means a small effect. The apparent reduction of magnetic moment by 12% means a 23% reduction in intensity.

A number of measurements have been made for the ion Fe^{3+} in different environments. In $LaFeO_3$ and $YFeO_3$, where the spin transfer is between Fe^{3+} and oxygen, it was found by Tofield and Fender[8] that the value of $A_\sigma^2 + 2A_\pi^2 + A_s^2$ was 10.0 ± 0.5% and 11.0 ± 1.0% in the

96 Measurements of Covalency

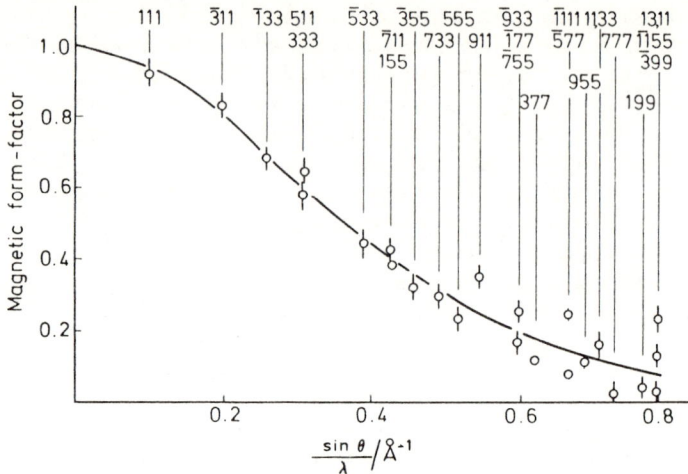

Figure 8.3 Experimental form-factor curve for NiO. The magnetic intensity data from a single-crystal, extrapolated to sin θ = 0, gave a value of 1.81 ± 0.20 μ_B for the magnetic moment of Ni^{2+}. The indices for the individual points are indicated at the top of the diagram (From Alperin[6], by courtesy of J. Phys. Soc. Japan)

two cases, respectively. On the other hand, in FeF_3 where the spin transfer is from the more electronegative fluorine ion, Jacobsen, McBride and Fender[9] find that the covalency is markedly lower, with a value of $A_\sigma^2 + 2A_\pi^2 + A_s^2$ equal to 6.2 ± 0.7%.

We turn now to examples of cases where there is a net transfer of unpaired spin density to the ligands, so that the third term in equation 8.2 is not zero. The best example is MnF_2, whose structure is illustrated in *Figure 8.4*, which was first studied with neutrons by Erickson[10] in 1953 but was re-examined, from the point of view of covalency, by Nathans et al.[11] ten years later. The magnetic cell is identical in size with the tetragonal chemical cell for which a = 4.87 Å and c = 3.31 Å. The figure shows that the environment of the fluorine ions is not symmetrical. Thus F(1) at the point *xyz* has two near-neighbours Mn(1) and Mn(2) whose magnetic spins point upwards and a third near-neighbour Mn(3) with a downward-pointing spin. Consequently F(1) gains a resultant upward-pointing spin and the same occurs for F(2); on the other hand, the other two fluorine ions in the unit cell will gain downward-pointing spins. If there had not been any transfer of spin to the fluorine ions then the magnetic structure-factor would be determined purely by the manganese atoms at 000 and $\frac{a}{2}\frac{a}{2}\frac{c}{2}$ and would be proportional to

$$fp_{Mn}[1 - \exp\{\pi i(h + k + l)\}]$$

and would be zero when $h + k + l$ had an even value. In fact, because of the spin transfers, there will be a magnetic contribution from the fluorine ions, amounting to $2fp_F \cos 2\pi hu \cos 2\pi ku$ when $h + k + l$ is odd and $2fp_F \sin 2\pi hu \sin 2\pi ku$ when $h + k + l$ is even. Here *u* is the

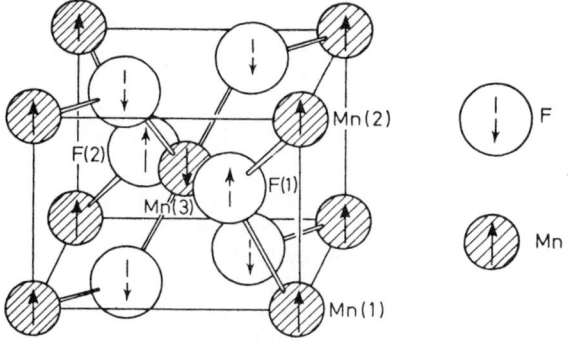

Figure 8.4 The magnetic structure of MnF$_2$, indicating the unsymmetrical environment of the fluorine ions, resulting in their retention of a net unpaired spin

parameter which defines the positions of the fluorine atoms and equals 0.31. It follows therefore that the effect of covalency will be to produce a magnetic contribution to those reflections for which $h + k + l$ is even and which would otherwise be purely nuclear. There will, however, be no contribution if either h or k is zero. The contributions will be relatively small, amounting in fact to a scattering length of 0.02×10^{-12} cm per atom in comparison with a value of 1.35×10^{-12} cm for an Mn^{2+} ion, but sufficiently large to be distinguished in measurements with polarized neutrons. Experimentally this means that the intensities for these normally forbidden reflections will *not* be the same for upward- and downward-pointing incident neutrons and from the ratio of the two intensities the magnetic contribution can be evaluated. The results of Nathans et al.[11] are shown in *Figure 8.5* and it will be

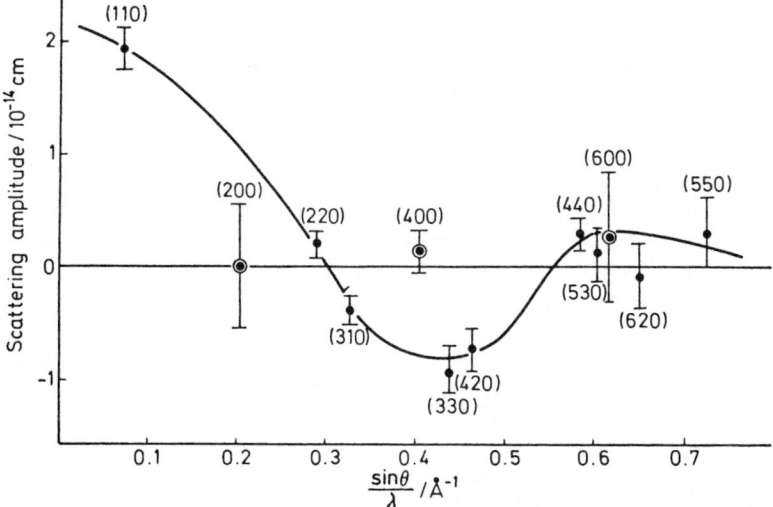

Figure 8.5 Measurements of the intensity of magnetic scattering in the 'forbidden' reflections of MnF$_2$, for which $h + k + l$ is even; intensity is not expected if h or k is zero (From Nathans et al.[11], by courtesy of J. Appl. Phys.)

noticed that although in general there is magnetic intensity for $h + k + l$ even, nevertheless this is zero as expected for the $h00$ reflections. From the magnitude of the forbidden magnetic intensities it is concluded that there is a magnetic moment of the order of 0.1 μ_B associated with each fluorine ion. If the contribution of the fluorine ions to *all* the magnetic reflections could be measured, then it would be possible to construct a Fourier plot of the spin density in the neighbourhood of the fluorine ion. Unfortunately the contribution from this density to those reflections for which $h + k + l$ is odd is swamped by the contribution from the manganese atoms, and cannot be assessed separately. However, from a partial Fourier synthesis it is possible to conclude that the covalent spin density is not concentrated at the centre of the fluorine ions but in regions indicated in *Figure 8.6*. Each Mn—F bond carries positive and negative clusters of spin density, with a resulting overall spin in the neighbourhood of the fluorine with a direction which is opposite to that on its two manganese neighbours Mn_A and Mn_B and of the same direction as that on Mn_C.

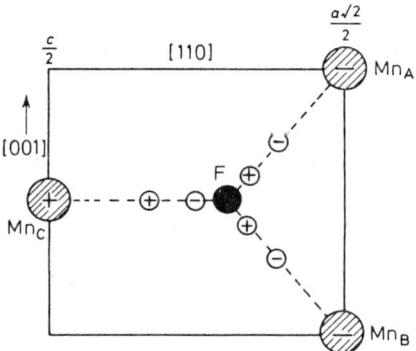

Figure 8.6 The distribution of covalent magnetic spin in the neighbourhood of the fluoride ions in MnF_2. *There is a net spin near F which is of opposite sign to that of the two parallel-oriented neighbours* Mn_A *and* Mn_B *(After Nathans et al.*[11]*, by courtesy of J. Appl. Phys.)*

Another instructive example is provided by $MnCO_3$, which has the rhombohedral structure, with space group $R\bar{3}c$, illustrated in *Figure 8.7*. The magnetic structure is predominantly antiferromagnetic, with the magnetic spins lying in the basal plane which is perpendicular to the trigonal axis and with the spins of the atoms at 000 and ½½½ oppositely directed. However, the two spin directions are not quite parallel, but slightly canted to produce a small ferromagnetic moment, which also lies in the basal plane. This resulting moment is only 0.04 μ_B per manganese atom, but this is large enough for measurement with polarized neutrons for a single crystal (Brown and Forsyth[12]). So far as this ferromagnetic moment is concerned there is a transfer, by covalency, of some of the spin density from the Mn^{2+} ions to the CO_3^{2-} anions. For this ferromagnetic component the spin density transfer to the ligands will, as we have shown earlier, *not* be cancelled out by effects from

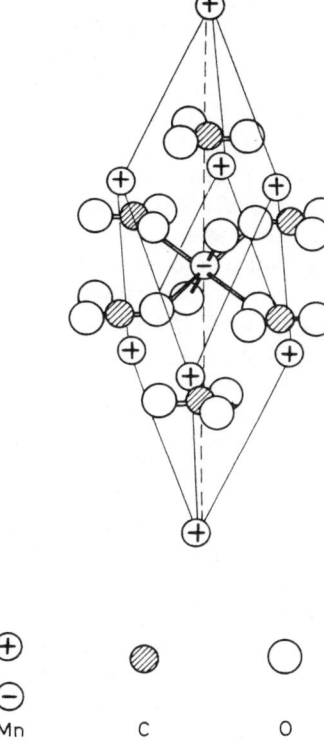

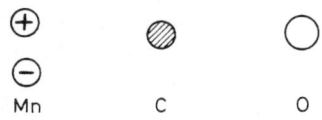

Figure 8.7 The structure of MnCO$_3$. The spin on the manganese atoms at the corners of the rhombohedral cell is almost antiparallel to that on the atoms at the body-centre, leaving a weak ferromagnetism

other surrounding cations. Thus we should expect the charge on the Mn^{2+} ions to be reduced and for a net spin density to be found on the anions. Consideration of the atomic positions in the structure leads to the conclusion that the ferromagnetic spin density, alone, contributes to reflections of the form hkl for which l is even. Thus Fourier maps of this spin density can be drawn. Later Lindgard and Marshall[13] made a more detailed examination of Brown and Forsyth's data, particularly of the high-field data in which a field of 7 kG applied in the [2$\bar{1}\bar{1}$] direction ensured that only domains of a single orientation were present, with no likelihood of the other two alternative domains which are crystallographically equivalent. This analysis concluded that the canted moment was equal to 0.08 μ_B and that covalency reduced this value by 8%, at the same time transferring an equal 8% to each CO$_3$ group, this latter being made up of 4% on each of the three oxygen ions and a partly compensating 4%, of opposite sign, on the carbon atom.

In those cases where the spin on the ligand is not cancelled by contributions of opposite sign from neighbouring cations, there will be a contribution to the form-factor, as indicated in *Figure 8.2* for the curve shown as 'paramagnetic'. As the figure shows, this contribution will be sharply peaked in angle, close to the forward position at $\theta = 0°$ and falling to zero near $Q = 1$ Å^{-1}. This happens because the cation-anion distance will be about 2 Å so that the distribution of spin over the magnetic complex will be quite extensive and diffuse, leading to a sharp peak. At first sight it would be expected that low-angle Bragg reflections would show evidence of this peak. However, a suitably placed peak would require a large interplanar spacing, measuring at least 10 Å. None of the well-understood common magnetic materials satisfy this condition and the effect has not yet been demonstrated directly in this way. The most convincing evidence so far comes in an experiment by Hirakawa and Ikeda[14], who have measured the paramagnetic scattering from the two-dimensional ferromagnet K_2CuF_4.

The measurements which we have described so far have established the essential principles of the effect of covalency on measurements of neutron scattering. Subsequent work has been much more concerned with the accuracy of determination of the covalency coefficients and the way in which these vary for different cations and anions. *Table 8.1* lists the magnetic-moment reductions which are expected for a number of different ions in octahedral complexes. For d^3 ions A_π^2 is determined directly but for d^5 and d^8 only the sums $A_\sigma^2 + 2A_\pi^2 + A_s^2$ and $A_\sigma^2 + A_s^2$ can be determined, respectively. The table also indicates the quantity which can be determined by resonance measurements, using NMR and ESR methods which give the spin on the ligands in cases for which the anions possess nuclear spin. Thus for single crystals of $KNiF_3$ the nuclear magnetic resonance of the ^{19}F nucleus, with spin ½, will be affected by the occupancy of the $p\sigma$ orbitals transferred by covalency (Hall et al.[15]). The NMR measurements are done at room temperature, in the paramagnetic region where the spin transfer to the ligands is not nullified by the symmetry of the environment which exists for antiferromagnetic order. $KNiF_3$ has a cubic perovskite structure in which the Ni^{2+} ion is surrounded by a regular octahedron of F^- ions, giving the complex $(NiF_6)^{4-}$. In a corresponding way ESR measurements can be made for ions such as Cr^{3+} and Mn^{2+} when introduced as impurity ions on the regular octahedral sites in cubic crystals, such as the Mg sites in $KMgF_3$. Thus, for example, the complex $(CrF_6)^{3-}$ can be examined. It follows from *Table 8.1* that for d^5 ions a joint use of the neutron and resonance data would give the individual values of A_σ and A_π, which would not otherwise be obtainable. We emphasize also that neutron measurements can readily be made with oxides, but resonance measurements can only be done with ^{17}O, since ^{16}O has a zero nuclear spin.

Comparison of neutron and resonance data has led to some important conclusions, although also revealing certain discrepancies, which are still not explained. The d^8 ions, represented by Ni^{2+}, were generally thought to be understood: thus the NMR result of $A_\sigma^2 = 3.8\%$ and $A_s^2 = 0.54\%$ for $KNiF_3$ was compatible with the neutron result of $A_\sigma^2 + A_s^2 = 2.6 \pm 1.8\%$. At the same time this agreed value for fluorine was consistent with the neutron result of $A_\sigma^2 + A_s^2 = 3.8\%$ for NiO. However, this apparent uniformity has been upset by a recent measurement by ENDOR

Table 8.1 Comparison of neutron and resonance covalency factors*

Ions	Electrons	Paired	Unpaired	Moment reduction (neutrons)	Ligand spin (resonance)
Cr^{3+}, Mn^{4+}	d^3		t_{2g}^3	$4A_\pi^2$	A_π^2
Mn^{2+}, Fe^{3+}	d^5		$t_{2g}^3 \, e_g^2$	$\frac{6}{5}(A_\sigma^2 + 2A_\pi^2 + A_s^2)$	A_s^2
					$A_\sigma^2 - A_\pi^2$
Ni^{2+}	d^8	t_{2g}^6	e_g^2	$3(A_\sigma^2 + A_s^2)$	A_σ^2
					A_s^2

*A^2 is the nomenclature of Hubbard and Marshall[1] and is equivalent to the f of Tofield[4]

for Ni^{2+} ions in ^{17}O-doped MgO which gives a value of $A_\sigma^2 = 8.5\%$ and $A_s^2 = 0.7\%$ (Freund[16]).

For the d^3 ions, represented by Cr^{3+} and Mn^{4+}, a marked discrepancy appeared between the neutron and resonance data, both of which are considered to measure A_π^2. Thus the neutron measurement of $A_\pi^2 = 0.9 \pm 0.4\%$ in CrF_3 (Jacobson, McBride and Fender[9]) was completely different from $A_\pi^2 = 4.9 \pm 0.15\%$ for a resonance measurement of Cr^{3+} in $KMgF_3$ (Hall et al.[15]). Likewise for the oxide complexes the neutron value of $A_\pi^2 = 2.2 \pm 0.6\%$ for $LaCrO_3$ (Tofield and Fender[8]) was at variance with $A_\pi^2 = 7.1 \pm 0.7\%$ for Cr^{3+} ions in ^{17}O-doped MgO. However, the inconsistencies in these two pairs of results have been explained (Jacobson, McBride and Fender[17] and Tofield[4]) as the result of spin polarization, due to which the neutron and resonance data do *not* measure the same quantities. In systems which contain unpaired electrons there is an exchange coupling which lowers the energy of an orbital with parallel spin relative to that of its pair with antiparallel spin. The difference is proportional to the number of unpaired electrons and is particularly large for the d^3 ion where small amounts of unpaired spin appears in the e_g σ-bonding orbitals, which are polarized parallel to the existing spin of the t_{2g} orbitals of the Cr^{3+} or Mn^{4+} ion. This results in an *increase* of the spin on the ligand from the normal A_π^2 to $A_\pi^2 + A_\sigma^2$ but a *decrease* in the moment reduction of the cation. The moment of the latter now falls only to $1 - 4A_\pi^2 + 2A_\sigma^2$ instead of to $1 - 4A_\pi^2$. With this interpretation the results given above for Cr^{3+} can be reconciled. Choosing $A_\sigma^2 = +2.7 \pm 0.4\%$ and $A_\pi^2 = 2.2 \pm 0.4\%$, we have

for resonance $\quad A_\pi^2 + A_\sigma^2 = 4.9\%$

for neutrons $\quad A_\pi^2 - \dfrac{A_\sigma^2}{2} = 0.9\%$

which are in agreement with the experimental data. In the absence of spin polarization the value of A_σ^2 would be zero.

The d^5 ions Mn^{2+} and Fe^{3+} have been studied very considerably but, especially for Mn^{2+}, they still offer some unsolved problems. Compared with the d^8 and d^3 ions the situation is complicated by the fact that both σ and π covalency contribute and it is not possible to determine the individual parameters A_σ^2 and A_π^2 from either neutron or resonance results alone. Many measurements have been made of the form-factor of Mn^{2+} and it was at one time considered that there was a contraction of the form-factor curve relative to that expected for a free ion, but this conclusion does not seem to be substantiated. Quantitative measurements show a steady increase in covalency from MnO^{17} ($3.6 \pm 0.5\%$) through α-MnS^7 ($7.0 \pm 0.3\%$) and α-$MnSe^{18}$ ($7.5 \pm 0.3\%$) to $MnTe^{19}$ ($9.8 \pm 0.5\%$), whereas there is little difference between MnF_2 (3.3%) and MnO. On the other hand, for Fe^{3+} the covalency in the oxide ($11.8 \pm 0.5\%$) is substantially greater than for the fluoride ($6.2 \pm 0.7\%$) and combination with resonance data leads to the conclusion that the σ-covalency is much greater for the oxide ($7.6 \pm 0.8\%$) than for the fluoride ($4.1 \pm 0.6\%$).

There remains some uncertainty concerning spin polarization for the d^5 ions, with the possibility that polarization occurs in the σ-bonding 4s orbitals. Unfortunately, unlike for the d^3 case, the spin polarization effect cannot be deduced directly since it both decreases the charge which appears on the ligands and decreases the moment-reduction on the cation. Thus both neutron and resonance data are affected in the same manner.

References

1. HUBBARD, J. and MARSHALL, W., *Proc. Phys. Soc.*, **86**, 561 (1965)
2. MARSHALL, W. and LOVESEY, S.W., *Theory of Thermal Neutron Scattering*, Clarendon Press, Oxford (1971)
3. OWEN, J. and THORNLEY, J.H.M., *Reports Prog. Phys.*, **29**, 675 (1966)
4. TOFIELD, B.C., *Structure and Bonding*, **21**, 1 (1975)
5. HUTCHINGS, M.T. and GUGGENHEIM, H.J., *J. Phys. C*, 1303 (1970)
6. ALPERIN, H.A., *J. Phys. Soc. Japan*, **17**, Suppl. BIII, 12 (1962)
7. FENDER, B.E.F., JACOBSON, A.J. and WEDGWOOD, F.A., *J. Chem. Phys.*, **48**, 990 (1968)
8. TOFIELD, B.C. and FENDER, B.E.F., *J. Phys. Chem. Solids*, **31**, 2741 (1970)
9. JACOBSON, A.J., MCBRIDE, L. and FENDER, B.E.F., *J. Phys. C*, **7**, 783 (1974)
10. ERICKSON, R.A., *Phys. Rev.*, **90**, 779 (1953)
11. NATHANS, R., ALPERIN, H.A., PICKART, S.J. and BROWN, P.J., *J. Appl. Phys.*, **34**, 1182 (1963)
12. BROWN, P.J. and FORSYTH, J.B., *Proc. Phys. Soc.*, **92**, 125 (1967)
13. LINDGARD, P.A. and MARSHALL, W., *J. Phys. C*, **2**, 276 (1969)
14. HIRAKAWA, K. and IKEDA, H., *J. Phys. Soc. Japan*, **35**, 1608 (1973); *Phys. Rev. Lett.*, **33**, 374 (1974)
15. HALL, T.P.P., HAYES, W., STEVENSON, R.W.H. and WILKENS, J.J., *J. Chem. Phys.*, **38**, 1977 (1963)
16. FREUND, P., *J. Phys. C*, **7**, L33 (1974)
17. JACOBSON, A.J., TOFIELD, B.C. and FENDER, B.E.F., *J. Phys. C*, 1615 (1973)
18. JACOBSON, A.J. and FENDER, B.E.F., *J. Chem. Phys.*, **52**, 4563 (1970)
19. FENDER, B.E.F. and COFFIN, P.S., unpublished results
20. JACOBSON, A.J. in *Chemical Applications of Thermal Neutron Scattering*, ed. B.T.M. Willis, Oxford University Press (1973)

9
DEFECTS AND NON-STOICHIOMETRY

We have already seen that the scattering from a perfect three-dimensional array of identical atoms takes the form of sharp well-defined Bragg peaks. Further, we have noted that when the scattering centres cease to be identical, as for example when different isotopes are present or when the scattering length depends on the spin alignment of neutron and nucleus, then the Bragg peaks are accompanied by incoherent background scattering. In the two cases just mentioned, of isotope and spin incoherence, the background scattering will be distributed isotropically in space. Other departures from regularity, such as atomic disorder or the presence of defects like vacancies or interstitial atoms or random variations of atomic position, will similarly produce incoherent scattering as part of the background of the diffraction pattern. It is usual to call this 'diffuse scattering'; if the irregularities are of nuclear size and randomly distributed then the diffuse scattering will be isotropic, but if they are grouped in pairs or larger assemblies with an extent equal to several wavelengths of radiation, then the diffuse scattering will depend on the angle of scattering 2θ.

Accordingly, for the total intensity of scattering at any particular angle we may write an expression which is the sum of two components: first, the Bragg scattering (which has a finite value only at positions which are very close to reciprocal-lattice points) and, secondly, the diffuse scattering, as

$$\frac{d\sigma}{d\Omega} = \psi\psi^*, \text{ where } \psi = \sum b_m \exp(i\mathbf{Q}\cdot\mathbf{r}_m)$$

$$= \sum_{m,m'=1}^{M} b_m b_{m'} \exp(i\mathbf{Q}\cdot\mathbf{r}_m - \mathbf{r}_{m'}) \tag{9.1}$$

where the summation is made over an assembly of M atoms with vector distances $\mathbf{r}_m$ from an arbitrary origin. This expression becomes

$$\sum_{n}^{N}\sum_{n'} (\bar{F})^2 \exp(i\mathbf{Q}\cdot\mathbf{r}_n - \mathbf{r}_{n'}) + \sum_{n}^{N}\sum_{n'} (F_n F_{n'} - \bar{F}^2) \exp(i\mathbf{Q}\cdot\mathbf{r}_n - \mathbf{r}_{n'}) \tag{9.2}$$

where the summation is made over N unit cells, which are not quite identical in content and where $\mathbf{r}_n$ defines the position of a unit cell relative to the arbitrary origin. F is the *average* value of the structure-amplitude factor for a unit cell and F_n is the *actual* value of the structure-amplitude factor for cell n. The two terms in equation 9.2 constitute the Bragg scattering and the diffuse scattering, respectively. It is evident that the Bragg scattering is determined by the average structure and the diffuse scattering by the departures from the average. Information about defects, i.e. the departures from regularity, will be forthcoming from each of these terms.

There is a very wide range of materials which may be regarded as containing 'defects' in this way, including solid solutions in which the ordering of the constituent atoms is not perfect, irradiated materials which contain interstitial atoms of various kinds, and non-stoichiometric compounds, in which at least a partial randomness of atomic distribution must be present. We shall examine these non-stoichiometric compounds in some detail and can classify them into three categories. First, when the concentration of defects is less than 1%, the defects are likely to be distributed singly and at random and free from any interaction amongst themselves. In these proportions they are unlikely to be detected by diffraction methods. On the other hand, proportions ranging from 1 to 10% will be found to produce observable effects on both the Bragg and diffuse scattering. Very often they are not distributed singly but in well-defined clusters which can be identified from the diffraction data. The third category covers those compounds such as the MoO_{3-x} system, which are not genuinely non-stoichiometric but contain localized distributions, at regular intervals, of ordered intermediate compounds.

Neutrons have several particular virtues for studying defect structures. Most important perhaps arises from the circumstance that most non-stoichiometric compounds contain light atoms, such as H, C, N and O, which are readily detected with neutrons but, in the presence of heavy atoms, are very difficult to locate with X-rays. At the same time many of these compounds are only stable at higher temperatures, where accurate intensity data are much easier to collect with neutrons than with X-rays and where the reduction of intensity of the high-angle reflections, through the Debye-Waller effect, is not so detrimental for neutrons, which show no angularly-dependent form-factor due to atomic size.

In using the Bragg scattering to elucidate the structure of a non-stoichiometric material it is convenient to express the average structure factor $\overline{F}$ in the form

$$\overline{F} = \sum_\kappa m_\kappa b_\kappa \exp(iQ \cdot r_\kappa) \exp(-W_\kappa Q) \qquad (9.3)$$

where the summation is made over the κ atomic positions at vectors r_κ, including the sites of interstitial atoms, in the unit cell and m_κ is an occupancy factor for each site. The final factor in the expression may be written alternatively in the form $\exp(-B_\kappa \sin^2\theta/\lambda^2)$, so that for each site there are three structural parameters m_κ, r_κ and B_κ to be determined by least-squares or Fourier methods.

The most complete investigations are those of compounds possessing the fluorite structure with an excess of anions and, as a particular example, the uranium oxide system at compositions such as $UO_{2.12}$. Single crystals were examined by Willis[1] for compositions ranging from UO_2 to U_4O_9 and over a temperature range from 20 to 1100 °C. It was found that the excess oxygen atoms do not occupy the large holes in the fluorite structure at positions ½½½, but are to be found on two types of interstitial site which are displaced by about 1 Å in the ⟨110⟩ and ⟨111⟩ directions respectively from these holes. At the same time a number of vacancies are found among the normal oxygen positions. At higher temperatures, both for UO_2 and UO_{2+x}, the oxygen atoms on the normal occupied sites are observed to relax slightly from the standard ¼¼¼ positions to ¼ + δ, ¼ + δ and ¼ + δ; δ increases steadily with temperature and equals 0.016 at 1200 K. The interstitial positions and their occupancy numbers, which for $UO_{2.12}$ are required to add up to a total of 2.12 atoms, are given in *Table 9.1*. It is further concluded, by

Table 9.1 Structural parameters of $UO_{2.12}$ in the average cell

Atom	Coordinates			Occupancy
	x	y	z	
Normal O	0.25	0.25	0.25	1.87 ± 0.03
Interstitial O′	0.5	0.38	0.38	0.08 ± 0.04
Interstitial O″	0.41	0.41	0.41	0.16 ± 0.06

considering the interatomic distances in terms of the known ionic radii, that the interstitial positions O′ and O″ and the vacancies in the O positions are *not* occupied at random, since this would bring several oxygen atoms too close together (Willis[2]), but in the form of 2 : 2 : 2 clusters. *Figure 9.1* shows how a pair of vacancies, on neighbouring sites, is accompanied by a pair of O′ interstitials (displaced along ⟨110⟩) and a pair of O″ interstitials (displaced along ⟨111⟩). On this assumption that the defects occur as 2 : 2 : 2 clusters the composition of $UO_{2.12}$ can be written as $UO_{1.88}O'_{0.12}O''_{0.12}$, which is within the accuracy of the experimental determination of the occupancy factors for the various sites. It is emphasized that the measurements of the intensities of the Bragg reflections determine the 'average unit cell'. Thus they reveal the occupied sites, and the percentage occupations, but of themselves do not give any direct information about the local distribution or of any correlation between the different kinds of defects.

The measurements with single crystals of U_4O_9 by Willis[1] showed it to have a unit cell which is four times as large as the basic UO_2 cell and to contain the same type of interstitial sites as $UO_{2.12}$. Effectively, the transition from UO_{2+x} to U_4O_9 involves long-range ordering of the defect complexes. However, more recent measurements by Masaki and Doi[3], to which we referred earlier in Chapter 4, suggest that the structure of U_4O_9 is more complicated than this and that displacements of the uranium atoms are also involved. The discrepancies between the two investigations may be considered to emphasize the difficulties in determining the details of this type of structure, with a multiple unit-cell and a very wide range of intensities of reflection. With single crystals the accuracy of intensity measurement depends on precise correction for extinction and any possible interpretation of powder data depends on a correct choice, or guess, of a model for which parameters can be refined.

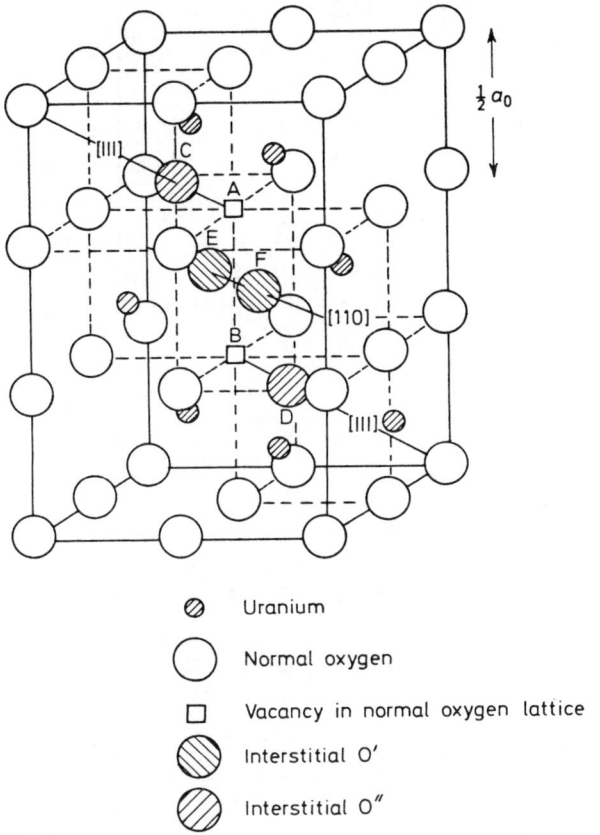

Figure 9.1 A model of the structure of UO_{2+x} to indicate the nature of the 2 : 2 : 2 complex. In UO_2 oxygen atoms would be present at A and B, whereas in UO_{2+x} they are replaced by a pair at C and D (displaced in the [111] direction) and a second pair at E and F (displaced in the [110] direction). These two pairs, together with the vacant sites A and B. constitute the 2 : 2 : 2 complex (After Willis[1], by courtesy of J. de Phys.)

As a further example of a fluorite structure with excess of anions we may consider the solid solutions of CaF_2 itself with up to 32% of YF_3, which were examined by Cheetham, Fender and Cooper[4]. These materials range in composition from CaF_2 to $(Ca/Y)F_{2.32}$ and were studied in both single-crystal and powder form. The single-crystal data, collected for $(Ca/Y)F_{2.06}$, permit the construction of Fourier projections of the scattering density. The general conclusion of this work is that interstitial fluorine atoms are present in the same type of position as was found above for the oxygen atoms in UO_{2+x} and, again, that vacancies occur in the ordinary fluorine positions. This is illustrated directly, for the 'average unit cell', in the difference Fourier projections on the (110) plane which are shown in *Figure 9.2*. In *Figure 9.2(a)* no interstitial atoms have been assumed; accordingly, positive scattering density should appear at interstitial sites and it is particularly evident in the region of

108   *Defects and Non-Stoichiometry*

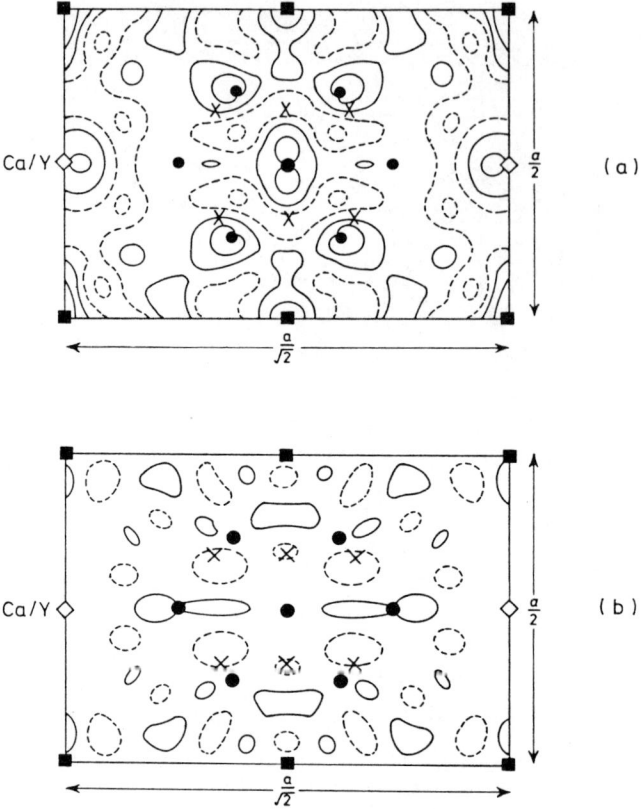

Figure 9.2 Difference Fourier projections on the 110 plane for single crystals of $(Ca/Y)F_{2.06}$. ■ *are normal F atoms,* ● *are the F' interstitial sites* ½vv, × *are the F'' interstitial sites www and* ◇ *are Ca/Y atoms.   (a) is a synthesis of $F_{obs} - F_{calc}$ for a model with no interstitial atoms; (b) is a synthesis for the final model of the defects (After Cheetham, Fender and Cooper[4], by courtesy of J. Phys. C)*

the ½uu, i.e. F', sites. On the other hand, in *Figure 9.2(b)* the difference function which has been plotted is derived by subtracting from the experimental F_{obs} a computed F_{calc} which takes account of all the interstitials and vacancies of the final model. This projection is relatively featureless and suggests that the 'average' model is largely correct. This use of the single-crystal data greatly helps to eliminate incorrect models. For each composition of material the powder data were refined by least-squares analysis to yield occupancy factors for the two types of interstitial site and for the normal fluorine sites, the latter occupancy being reduced below two by the vacancies. Results at room temperature for $(Ca/Y)F_{2.15}$, $(Ca/Y)_{2.25}$ and $(Ca/Y)_{2.32}$ are given in *Table 9.2*. From these conclusions about the average cell it is then possible to make suggestions concerning the local correlation of vacancies and interstitial atoms. In order to avoid very short F···F distances it is necessary to assume that vacancies and interstitials are associated in some form of cluster. When the anion excess x is equal to 0.06 or 0.10 it is possible

Table 9.2 Structural parameters for $(Ca/Y)F_{2+x}$ at room temperature in the 'average' cell

Atom	Value of x	Coordinates			Occupancy
Normal F	0.15	0.25	0.25	0.25	1.81 ± 0.02
	0.25	0.25	0.25	0.25	1.62 ± 0.02
	0.32	0.25	0.25	0.25	1.54 ± 0.02
Interstitial F'	0.15	0.5	0.36	0.36	0.21 ± 0.03
	0.25	0.5	0.38	0.38	0.47 ± 0.02
	0.32	0.5	0.37	0.37	0.48 ± 0.04
Interstitial F''	0.15	0.41	0.41	0.41	0.13 ± 0.04
	0.25	0.36	0.36	0.36	0.16 ± 0.03
	0.32	0.41	0.41	0.41	0.30 ± 0.04

to satisfy the observed values of the occupancy factors if all the defects are present as 2 : 2 : 2 clusters like those which we illustrated in *Figure 9.1* for UO_{2+x} and one of which is drawn from a different viewpoint in *Figure 9.3(a)*. At $x = 0.15$ agreement is better for the larger 3 : 4 : 2 cluster, which is drawn in *Figure 9.3(b)*. For compositions with $x = 0.25$ and 0.32 the data strongly suggest that the number of vacancies is significantly larger than the number of excess fluoride ions. This is not compatible with either the 2 : 2 : 2 or the 3 : 4 : 2 type of cluster and it was suggested that several of the latter type may aggregate to give larger clusters, such as the octahedral complex shown in *Figure 9.3(c)*. At the intersection of the six chains a new vacancy-interstitial pair has been created.

This interpretation of the defects in $(Ca/Y)F_{2+x}$ has been pursued further by Steele, Childs and Fender[5], who have examined the *diffuse* scattering from the same materials. Measurements of diffuse scattering can be made most readily if the experiments are carried out under such conditions that Bragg scattering does not take place, and this can be achieved by using a wavelength which is greater than twice the maximum interplanar spacing in the crystal. For CaF_2 the maximum spacing, d_{111}, is equal to 3.2 Å and a wavelength of 7 Å was used, with a time-of-flight analysis to remove all inelastically-scattered neutrons from the observed count. As we have already seen, the diffuse scattering will indicate, in a summation over the whole of the sample, the departures of the local atoms, interstitials and vacancies from the situation in the average unit cell. If the defects consist of random isolated interstitial atoms or vacancies, then the diffuse scattering will be very simple to evaluate, for it will amount to an isotropic differential cross-section

$$\frac{d\sigma}{d\Omega} = c(1-c)b_A^2 \qquad (9.4)$$

where c is the fractional concentration of interstitial atoms or vacancies and b_A is their scattering length. If the defects exist as clusters, then the scattering will cease to be isotropic and will have a form-factor type of dependence on the shape of the cluster, bearing in mind that the

110 Defects and Non-Stoichiometry

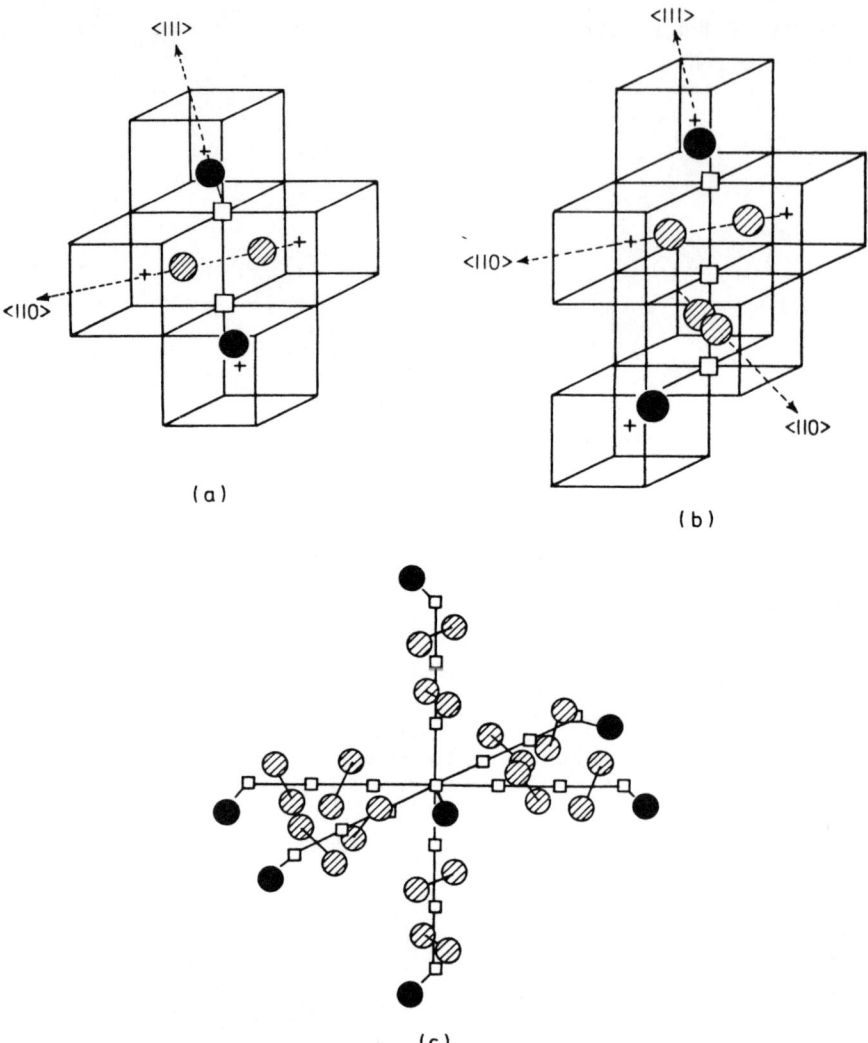

Figure 9.3 Cluster models in $(Ca/Y)F_{2+x}$. In (a) and (b) which, respectively, show 2 : 2 : 2 and 3 : 4 : 2 clusters, the unmarked corners of the large cubes are occupied by normal fluorine atoms. □ are normal fluorine vacancies, and ⊝● are types F' and F'' of interstitial atoms. The crosses + indicate the cubic holes such as ½½½. In (c), which shows a possible aggregated cluster, drawn on a smaller scale, the large cubes of normal fluorine atoms have been omitted (After Cheetham, Fender and Cooper[4], by courtesy of J. Phys. C)

cluster may take up any one of several orientations relative to the crystal axes. If it may be assumed that the individual clusters are themselves distributed randomly, then the diffuse scattering may be calculated in a straightforward manner for any assumed model of a cluster. In principle the observed scattering pattern can be Fourier-inverted to derive

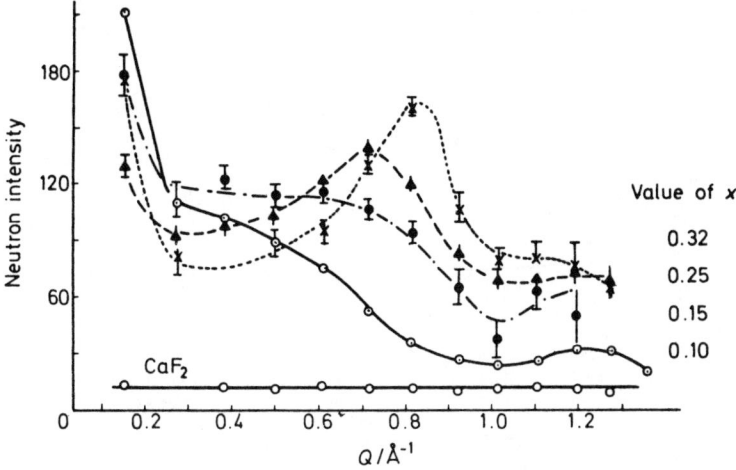

Figure 9.4 Diffuse scattering patterns for CaF_2 and $(Ca/Y)F_{2+x}$, for various values of x, measured with long-wavelength neutrons (After Steele, Childs and Fender[5], by courtesy of J. Phys. C)

the shape of the cluster, but this is not of much practical value when scattering data are only available over a restricted range of Q.

The diffuse scattering patterns for $(Ca/Y)F_{2+x}$ are shown in *Figure 9.4* for values of x from 0.1 to 0.32, together with the pattern for CaF_2 itself. For the latter the scattering is very small and isotropic, as expected, arising from the small isotopic- and spin-incoherence of calcium and fluorine. It is noteworthy in the figure that for $x = 0.25$ there is a broad maximum in the scattering pattern with a peak around $Q = 0.7$ Å^{-1}; for $x = 0.32$ the peak is sharper and has moved to 0.8 Å^{-1}. The authors have compared these curves with calculations for a wide variety of alternative models and conclude that none provides a better description of the scattering than the forms of cluster which, as we saw above, had been suggested earlier to account for the details of the Bragg scattering. The different assumed models take account of the fact that because of the overall negative charge associated with a cluster this charge will have to be neutralized by Y^{3+} ions, which could be associated with the cluster in various ways. It appears that there is only a rather loose association and agreement with experiment is less good if it is assumed that the yttrium ions specifically occupy the cation sites closest to the cluster. *Figure 9.5* compares some computed and observed curves for $(Ca/Y)F_{2.1}$ and $(Ca/Y)F_{2.25}$. In the former the experimental points are shown in comparison with sets of curves for 3 : 4 : 2 and 2 : 2 : 2 clusters, respectively. In each case curves are shown in turn for (i) a random distribution of Y^{3+} ions, (ii) for Y^{3+} at 000, ½½0 and ½0½ and (iii) for Y^{3+} at 0½½, 111 and 10$\bar{1}$. The latter assumption, with the Y less closely associated with the cluster, gives the best agreement. For $(Ca/Y)F_{2.25}$ the computed curves show how the six-fold multicluster which we illustrated in *Figure 9.3(c)* will produce a peak in the neighbourhood of $Q = 0.7$ Å^{-1}. The two further curves in this figure,

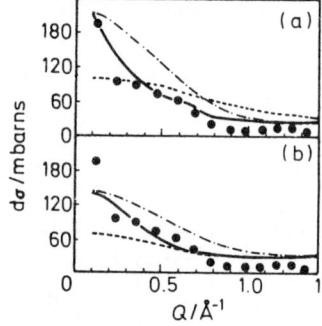
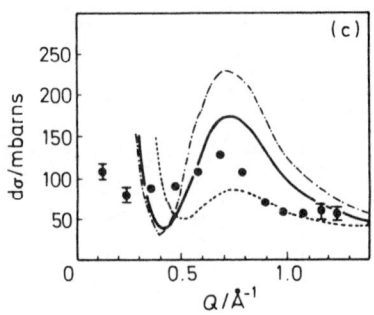

Figure 9.5 Comparison of experiment and calculation for the diffuse scattering of $(Ca/Y)F_{2.1}$ (a), (b) and $(Ca/Y)F_{2.25}$ (c); ● are the experimental points. In (a) the calculated curves are for a 3 : 4 : 2 cluster and in (b) for a 2 : 2 : 2 cluster. In each case ----- is calculated for a random distribution of Y^{3+}, —·—·—· is for a close association with the cluster, at 000, ½½0 and ½0½; ——— is for a looser association, at 0½½, 111 and 10$\overline{1}$. In (c) ——— is calculated for the multi-cluster of Figure 9.3(c); ..... also includes four central F' ions, and —·—·—· is without any central interstitial ion, in each case for a random distribution of Y ions (After Steele, Childs and Fender[5], by courtesy of J. Phys. C)

which give less satisfactory agreement, are for modified six-fold clusters with, respectively, four central F' interstitial ions and no central ion.

Among the many compounds with the rock-salt type of structure which have been examined, the iron-deficient FeO is probably the best known. This is usually found in compositions between $Fe_{0.91}O$ and $Fe_{0.95}O$ and is of interest both from chemical and magnetic points of view. FeO is antiferromagnetic below 198 K and Roth[6] showed that only about 75% of the normal cation sites in the structure were occupied, with a substantial number of iron atoms in tetrahedral interstitial sites. He also showed that many of the magnetic spins did not take part in the antiferromagnetic ordering and postulated the existence of paramagnetic islands in the material, within which two octahedral vacancies and one tetrahedral interstitial iron atom were associated. Cheetham, Fender and Taylor[7] attempted to clarify this situation by making neutron measurements between 1070 and 1470 K, knowing that FeO is only in stable equilibrium above 840 K and the earlier work had been done with quenched samples. They concluded that the occupation of tetrahedral sites by iron atoms was about 3%, and that although there undoubtedly were clustered defects it seemed likely that these contained vacancies and interstitial atoms in a ratio between 3 and 4, and certainly greater than the value of 2 suggested by Roth and possibly larger than the 13 : 4 ratio for the type of cluster proposed by Koch and Cohen[8] in an X-ray investigation.

The non-stoichiometric compound V_8C_7 provides another interesting example of the value of using neutrons. It had been suggested by de Novion, Lorenzelli and Costa[9] from X-ray data that the unit cell of this compound was based on a doubled version of the rock-salt cell which would be appropriate to VC. Such a cell would normally contain 32 atoms of vanadium and 32 of carbon and it was postulated that 4 of the latter were absent, giving an idealized composition $VC_{0.875}$, and that the vacancies were ordered in a helical spiral to account for the observed X-ray intensities. For neutrons the coherent scattering amplitude of vanadium is very small indeed, being only -0.05×10^{-12} cm in comparison with 0.66×10^{-12} cm for carbon, so that the intensity in the Bragg reflections is contributed almost entirely by the carbon atoms. Accordingly the intensities of the Bragg reflections will provide a very good test of the suggested spiral-ordering of the vacancies. Measurements were made by Henfrey and Fender[10] at 4 and 300 K for $VC_{0.863}$ and the occupancy numbers for the three types of carbon site were in good agreement with what is expected for a perfect spiral array. At the same time it was found that the assumption that the carbon atoms were not displaced from their normal positions in the fluorite cell was justified by the low values of apparent temperature-factors which resulted.

As further examples of systems for which the diffuse component of the scattering has been studied we mention the carbon-deficient rock-salt structure of $NbC_{0.86}$ and the interstitial atoms of hydrogen which can be taken up by vanadium and niobium. $NbC_{0.86}$, which is typical of many quasi-metallic compounds MX_{1-x} which are formed by Ti, V and Nb with C, N and O, has been studied by Henfrey and Fender[11] and the diffuse scattering pattern, shown in *Figure 9.6*, makes it very clear that the vacancies are not randomly distributed. The curve in the figure, which reproduces the peak at $Q = 1.4$ Å^{-1} found experimentally, is calculated for a model in which the occupancy by carbon atoms of the first three coordination shells around a vacancy is assumed to be 0.96, 0.92 and 0.79, respectively. For a random distribution of vacancies each probability would be 0.86 so that there is a vacancy-vacancy repulsion which requires the vacancies to be disproportionately surrounded by carbon atoms. On the other hand, the interstitial atoms in $NbD_{0.079}$, in which the 8% of deuterium is dissolved in the body-centred structure of the metal, behave quite differently[12]. It would be expected that the deuterium atoms would

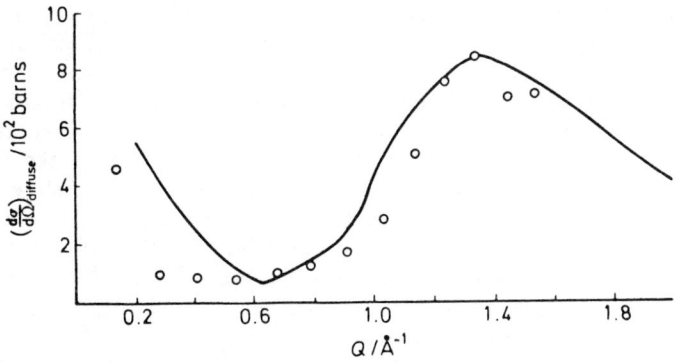

Figure 9.6 The diffuse scattering by $NbC_{0.86}$. ○ are the experimental points and the curve is calculated for a model in which vacancies are surrounded by an excess of carbon atoms in first and second coordination shells, indicating vacancy-vacancy repulsion (From Henfrey and Fender[11])

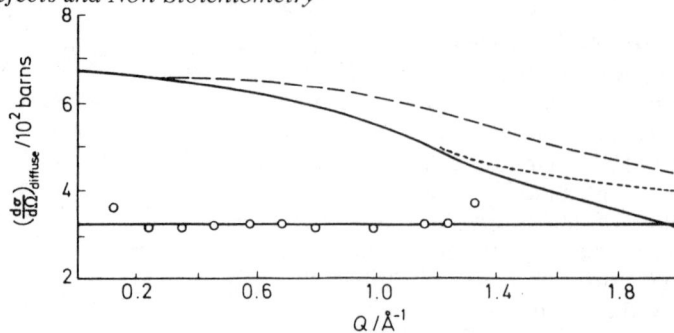

Figure 9.7 Variation of the diffuse scattering by $NbD_{0.079}$. ○ are the experimental points and the horizontal straight line through them is calculated for a random distribution of isolated deuterium atoms. —— is calculated for random pairs of D atoms on next-nearest-neighbour tetrahedral sites such as ½0¼ and ½0¾; ···· is calculated for random pairs on near-neighbour tetrahedral sites such as ½0¼ and ¾0½; --- is calculated for random pairs on near-neighbour tetrahedral sites, accompanied by displacement of neighbouring niobium atoms by 0.1 Å in the [111] direction (From Fender and Henfrey[12], by courtesy of J. Chem. Phys.)

be found in the tetrahedral holes such as ½0¼ and previous thermodynamic evidence had suggested that they would form clusters. However, the diffuse scattering pattern for neutrons of wavelength 4 Å, shown in *Figure 9.7*, is practically isotropic, suggesting that the interstitial atoms occur singly, at random. Any clustering would produce a marked concentration of scattering in the forward direction, as indicated by the curves in the figure which were calculated for various suggested models of the cluster. The use of deuterium, rather than hydrogen, for these experiments is to ensure that the relatively small effects being sought in the diffuse scattering would not be swamped by the enormous spin-incoherent scattering of ordinary hydrogen.

Two other recent studies of the niobium–deuterium system have been made, one in the region of 3% of deuterium and the other near the equiatomic composition. In the latter case Somenkov et al.[13] examined powdered material with the composition $NbD_{0.95}$. Above 400 K a completely disordered state with a body-centred cubic structure of side a_0 is found, but below this temperature there is a transition to a tetragonal structure with sides approximately $a_0\sqrt{2}$ and a_0 and space group *Pnnn* in which the deuterium atoms are ordered in tetrahedral interstices. In studies of single crystals of $NbD_{0.03}$, Somenkov et al.[14] find that at room temperature there is a disordered solid solution, showing only the peaks of the body-centred-cubic metal, but additional weak peaks appear below 270 K which are believed to be due to precipitation of $NbD_{0.70}$. The latter has the same basic structure as was found for $NbD_{0.95}$, but only about three-quarters of the deuterium sites can be occupied. On further cooling, below 170 K, extremely weak further peaks appear, owing to additional ordering which was explored in detail with a separate powder sample of composition $NbD_{0.755}$. The ordered phase was ascribed to the orthorhombic space group $P2_1 2_1 2_1$ with a unit cell for which a and b are approximately equal to $2a_0$ and c is approximately the same as a_0, the unit-cell edge for niobium metal.

References

1. WILLIS, B.T.M., *J. de Phys.*, **25**, 431 (1964)
2. WILLIS, B.T.M., *Proc. Brit. Ceram. Soc.*, **1**, 9 (1964)
3. MASAKI, N. and DOI, K., *Acta Cryst.*, **B28**, 785 (1972)
4. CHEETHAM, A.K., FENDER, B.E.F. and COOPER, M.J., *J. Phys. C*, **4**, 3107 (1971)
5. STEELE, D., CHILDS, P.E. and FENDER, B.E.F., *J. Phys. C*, **5**, 2677 (1972)
6. ROTH, W.L., *Acta Cryst.*, **13**, 140 (1960)
7. CHEETHAM, A.K., FENDER, B.E.F. and TAYLOR, R.I., *J. Phys. C*, **4**, 2160 (1971)
8. KOCH, F. and COHEN, J.B., *Acta Cryst.*, **B25**, 275 (1969)
9. DE NOVION, C.H., LORENZELLI, R. and COSTA, P., *C.R. Acad. Sci. Paris*, **263B**, 775 (1966)
10. HENFREY, A.W. and FENDER, B.E.F., *Acta Cryst.*, **B26**, 1882 (1970)
11. HENFREY, A.W. and FENDER, B.E.F., Unpublished results
12. FENDER, B.E.F. and HENFREY, A.W., *J. Chem. Phys.*, **52**, 3250 (1970)
13. SOMENKOV, V.A., ZEMLYANOV, M.G., KOST, M.E., CHERNOPLEKOV, N.A. and CHERTKOV, A.A., *Sov. Phys.-Dokl.*, **13**, 669 (1969)
14. SOMENKOV, V.A., PETRUNIN, V.F., SHIL'SHTEIN, S.Sh. and CHERTKOV, A.A., *Sov. Phys.-Cryst.*, **14**, 522 (1970)

10

MOLECULAR SPECTROSCOPY

10.1 Principles of inelastic incoherent scattering

We have indicated in Chapter 3 that during the process of scattering a neutron will undergo a change in momentum and often in energy. If no change in energy occurs then the scattering is 'elastic'; otherwise it will be inelastic. The part of the elastic scattering for which there are defined phase relations between neighbouring atoms, determined by their spatial positions, is said to be 'coherent'. Incoherent elastic scattering will occur when materials contain isotopes of different scattering length or when there are nuclei with spin-dependent scattering lengths. For a single crystal we can express the scattering as a differential cross section, rather than solely in terms of a scattering amplitude and a structure amplitude factor, as

$$\left(\frac{d\sigma}{d\Omega}\right)_{coh} = \frac{N_0}{v} \sum_{\tau} \delta(Q - \tau) F_{\tau}^2 \qquad (10.1)$$

where v is the volume of the unit cell and N_0 is the number of unit cells in the sample. This expression is the cross-section per unit solid angle for a momentum transfer vector Q; the occurrence of the δ function indicates that for a crystal containing many unit cells, intensity will only be observed if Q coincides with a reciprocal lattice vector τ, thus accounting for the characteristic sharp Bragg reflections. In the expression F_{τ} is the unit cell structure-factor, given by

$$F_{\tau} = \sum (\bar{b}) \exp i(\tau \cdot \rho) \qquad (10.2)$$

where ρ is an atomic position vector and $\bar{b}$ is the coherent scattering amplitude. On the other hand, the analogous contribution to incoherent scattering is

$$\left(\frac{d\sigma}{d\Omega}\right)_{incoh} = \sum \left\{ (\overline{b^2}) - (\bar{b})^2 \right\} = \sum b_{incoh}^2 \qquad (10.3)$$

summed over all the nuclei, and this will yield isotropic scattering, independent of angle.

In a similar way, cross-sections can be defined for the inelastic scattering, but in this case the differential cross-section at any angle of scattering must be further subdivided according to the energy E of the scattered neutrons, thus leading to the double differential cross-section per unit of energy range near E, specified as $d^2\sigma/d\Omega dE$. This will be related in a very general way to the dynamics of the solid, rather than simply to the static positions which are assessed by the elastic scattering, and in principle will be defined by both the positions and velocities of all the atoms. Thus

$$\frac{d^2\sigma}{d\Omega dE} = N \frac{\kappa}{\kappa_0} \frac{1}{2\pi\hbar} \int_{-\infty}^{\infty} dt \exp(-i\omega t) \sum_r \exp(i\mathbf{Q}\cdot\mathbf{r})$$

$$\times \{(\bar{b})^2 G(r,t) + [\overline{b^2} - (\bar{b})^2] G_s(r,t)\} \qquad (10.4)$$

where N is the number of scattering nuclei and $G(r,t)$ and $G_s(r,t)$ are the space-time correlation functions of Van Hove[1]. The transfer of energy is $E_0 - E = \hbar\omega$. The two portions of this expression are termed the coherent and incoherent inelastic scattering cross-sections. Thus

$$\left(\frac{d^2\sigma}{d\Omega dE}\right)_{coh} = N \frac{\kappa}{\kappa_0} \frac{1}{2\pi\hbar} \int_{-\infty}^{\infty} dr\, dt \exp i(\mathbf{Q}\cdot\mathbf{r} - \omega t)(\bar{b})^2 G(r,t) \qquad (10.5)$$

and

$$\left(\frac{d^2\sigma}{d\Omega dE}\right)_{incoh} = N \frac{\kappa}{\kappa_0} \frac{1}{2\pi\hbar} \int_{-\infty}^{\infty} dr\, dt \exp i(\mathbf{Q}\cdot\mathbf{r} - \omega t)[\overline{b^2} - (\bar{b})^2] G_s(r,t) \qquad (10.6)$$

which, in turn, may be written as

$$\left(\frac{d^2\sigma}{d\Omega d\omega}\right)_{coh} = N \frac{\kappa}{\kappa_0} (\bar{b})^2 S(Q,\omega) \qquad (10.7)$$

and

$$\left(\frac{d^2\sigma}{d\Omega d\omega}\right)_{incoh} = N \frac{\kappa}{\kappa_0} [\overline{b^2} - (\bar{b})^2] S_s(Q,\omega) \qquad (10.8)$$

where the functions $S(Q,\omega)$ and $S_s(Q,\omega)$ are known as the coherent and incoherent scattering laws of the system. They depend on the dynamics of the scattering system and are independent of the properties of the scattered neutrons.

In the case of liquids, $S(Q,\omega)$ and $S_s(Q,\omega)$ provide the main method of describing the dynamics of the system. For crystalline solids it is usual to describe the motions of the atoms in terms of the various vibrations which may be present. Thus in a molecular crystal such as anthracene the vibrations can be considered to consist of a number of internal modes, whereby bonds may bend and stretch but individual molecules are regarded as independent, and a series of weaker external modes in which the molecules vibrate as rigid units. It turns out that the 'coherent' and 'incoherent' portions of the inelastic scattering reveal different features of the overall spectrum of vibrations which exist, or can be excited, in the solid. For a theoretical discussion of this conclusion the reader is referred to the book by Marshall and Lovesey[2]. The coherent scattering can reveal the dispersion law for the crystal relating the energy $\hbar\omega$ with the wavevector q for a phonon travelling in any chosen direction and for a vibration polarized in some specified direction. If we use a beam of monochromatic neutrons, i.e. of fixed κ_0, and examine the neutrons scattered by a single crystal in some chosen direction, we shall find that only neutrons which have undergone energy of exchange with some particular phonon q will be observed. This restriction exists because both momentum and energy equations (3.1 and 3.2) have to be satisfied. Thus by making measurements over a wide range of scattering angle we should be able to plot out the phonon spectrum. On the other hand, for the *incoherent* scattering there is no momentum conservation condition to be satisfied and in *any* direction of scattering neutrons of *all* wave-vectors and energies may be found. The scattered neutrons will have a continuous distribution in energy, from which the distribution function of vibrations $Z(\omega)$ can be extracted.

In practice two particular experimental procedures are attractive and informative. For substances of simple structure, like the metals and alkali halides and which contain elements whose scattering is almost entirely *coherent*, the study of single crystals will reveal the details of the various branches of the dispersion curves. Such studies have been made for many materials, leading directly to knowledge of the interatomic forces between atoms. The measurements can be made sufficiently accurately to give knowledge not only of the forces between neighbouring atoms but also of those out to about the tenth shell. Some measurements have been made of the dispersion curves in non-ionic molecular crystals, including deuterated hexamethylenetetramine (Dolling and Powell[3]), deuterated anthracene and deuterated *p*-dichlorobenzene (Reynolds, Kjems and White[4]). The latter was chosen because it is a favourable case with a rigid molecule and a simple structure, so that the intermolecular lattice modes are not obscured by other forces. Reasonable agreement was obtained between the experimental dispersion curves and those calculated from parameters assessed from such properties as the sublimation energy. At the other extreme, materials whose scattering is almost entirely *incoherent*, for example those containing hydrogen, can be examined in the polycrystalline, or powdered, form to yield the distribution function of vibrations $Z(\omega)$. Such a study of *incoherent inelastic* scattering provides a form of neutron spectroscopy which can cover a very wide range of frequencies, encompassing both the optical and acoustical phonons and the variety of intramolecular modes. It will be important to see how this type of neutron molecular-spectroscopy can provide information which is not forthcoming from, say, infrared or Raman spectra and, in particular, the way in which it can lead to the identification of vibrations and not simply make a measurement of their frequencies. It will be possible to recognize the main principles of neutron spectroscopy by considering first the equation which relates the incoherent scattering cross-section $(d^2\sigma/d\Omega dE)_{incoh}$ with the density of states distribution function $Z(\omega)$. As deduced by Marshall and Lovesey[2] and discussed, for example, by Lomer and Low[5], Windsor[6] and White[7],

$$\left(\frac{d^2\sigma}{d\Omega dE}\right)_{incoh} = \frac{\kappa}{\kappa_0} \sum_\nu \{(b_\nu^2) - (\bar{b}_\nu)^2\} \frac{Q^2 U_\nu^2}{2M_\nu} \exp(-2W_\nu)$$

$$\times \left\{\frac{1}{\exp(\epsilon/kT) - 1} + \frac{1}{2} \pm \frac{1}{2}\right\} N \frac{Z(\omega)}{\omega} \qquad (10.9)$$

for a sample containing N atoms, where the summation is made over the ν atoms in the molecule and the disposable sign is taken as + or − according as the neutrons lose or gain energy. We recall that κ_0 and κ are the wave-vectors, $2\pi/\lambda$, of the incident and scattered neutrons, Q is the momentum transfer vector so that $Q = \kappa - \kappa_0$, and the change of neutron energy ϵ is equal to $E - E_0$, or to $\hbar\omega$ with ω an angular frequency. Strictly, this expression has to be extended to include multiphonon terms but it is a good approximation in its simpler one-phonon form. It is evident from the form of this expression that there is not a direct correlation between the neutron spectrum and a *single* frequency-distribution function, but that each vibration is weighted by the different atoms taking part in it according to their values of $(b_\nu^2) - (\bar{b}_\nu)^2$, which can be written as b^2_{incoh}, and of U_ν, the atomic displacement in the relevant vibration, and the reciprocal of the atomic mass M_ν. The practical result of this weighting is that the neutron spectra are very substantially biassed in favour of showing up vibrations in which hydrogen atoms take part. Hydrogen is clearly favoured through its enormous incoherent scattering, its large amplitudes of vibration and its small mass. In a sense this may be regarded as a kind of 'selection rule' for neutron spectroscopy, but it must be emphasized that the selection rules applicable to IR and Raman spectroscopy, whereby only those vibrations which involve respectively a change in dipole moment or a change in polarizability are revealed, do not apply for neutrons. We also comment that when materials are deuterated there will be an immense reduction in intensity of those peaks in the neutron spectrum which involve the motion of hydrogen atoms. In IR spectroscopy, on the other hand, substitution of deuterium leads only to a change of vibration frequency. This change will depend on the amount of hydrogen motion in the vibration and may range from a few units of cm^{-1} to a reduction of frequency by a factor of $2^{1/2}$. Essentially it is the way in which IR, Raman and neutron spectroscopy are influenced by quite different features within molecules which lead to such immense power when the different techniques are used jointly.

In a practical experiment we have to consider the precise relationship between the number of neutrons recorded, at a given angle of scattering and with a given energy, and the quantity $d^2\sigma/d\Omega dE$ of equation 10.9. If, for example, a beryllium-filter detector is used with, in turn, incident beams of various wavelengths, then κ will be constant and negligibly small in comparison with κ_0. Hence $Q^2 = \kappa_0^2$ and $\omega = -\hbar\kappa_0^2/2m$. Thus Q^2/ω remains constant. At the same time the period of counting will normally be controlled by a monitor which assesses the incident beam. This monitor will have a sensitivity which is proportional to λ so that the number of neutrons it permits through for each counting point will be proportional to $1/\lambda$, i.e. to κ_0. This factor will cancel out the dependence on the κ_0 which is present in the denominator of equation 10.9. In this case, therefore, the variation with incident energy of the number of neutrons recorded will depend solely on

$$\sum_\nu (b_\nu^2)_{\text{incoh}} \frac{U_\nu^2}{2M_\nu} \left[\frac{1}{\exp(\epsilon/kT) - 1} + 1 \right] g(\omega) \exp(-2W_\nu) \qquad (10.10)$$

remembering that the neutrons are *losing* energy.

10.2 Typical applications

We shall proceed to give some examples of practical studies of molecular vibrations by neutron spectroscopy. In assessing some of the results the reader may find it useful to consult *Appendix 3*, on p. 180 at the end of the book, which gives a table of corresponding values of neutron wavelength λ in Å and neutron energy in electron-volts, together with the optical wavenumber, in cm^{-1}, equivalent to an energy exchange of this same amount.

In many very strongly hydrogen-bonded materials the bending vibration of the hydrogen bond gives an extremely broad band which is difficult to locate by infrared. With neutrons, however, the bending vibration is the most intense band in the spectrum and is very easily found. KHF$_2$, with its very strong hydrogen bond and the linear symmetric HF$_2^-$ ion, has been studied by Boutin, Safford and Brajovic[8] and by Collins, Haywood and Stirling[9] and a spectrum obtained by the latter is shown in *Figure 10.1*. The main peak at 1240 cm^{-1} is the fundamental bending vibration of the HF$_2^-$ ion, which has a large amplitude and involves motion of the hydrogen atom. On the other hand, the stretching vibration of the hydrogen bond at 600 cm^{-1} is very weak because it does not involve any hydrogen motion and is accounted for by the relatively small scattering from the fluorine atoms. The antisymmetric stretching mode is observed at 1473 cm^{-1} in the infrared, but it is a very broad line which for neutrons would be expected to have an overall intensity only one quarter of the bending mode and this accounts for its absence from the neutron pattern.

Evans and Lo[10] studied the IR and Raman spectra of a series of salts containing the Cl–H–Cl ion or its deuterated form Cl–D–Cl,

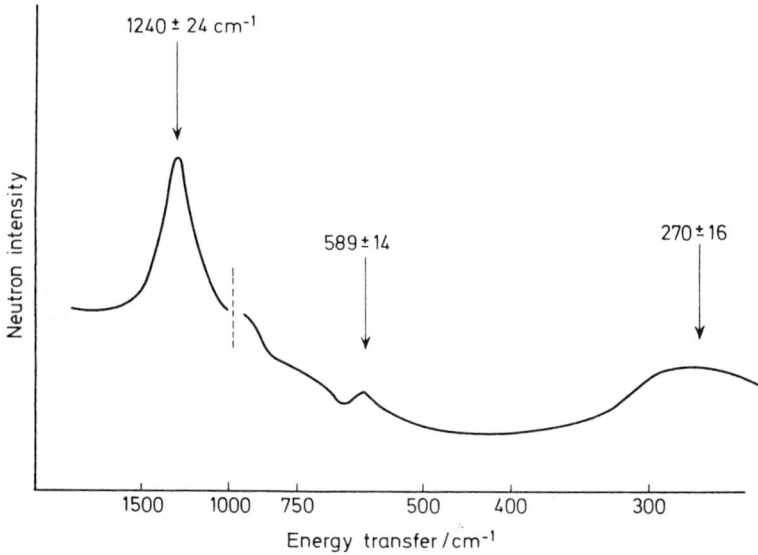

Figure 10.1 Energy-loss spectrum of KHF_2 using a beryllium-filter detector. The bending mode of the hydrogen bond gives an intense peak at 1240 cm^{-1}, but the symmetric stretch at 589 cm^{-1} is very weak because there is no hydrogen motion involved (After Collins, Haywood and Stirling[9], by courtesy of J. Chem. Phys.)

mainly of the form $R_4N[Cl-H-Cl]$, where R is an alkyl group, but also including the cesium salt $Cs[Cl-H-Cl]$. The spectra suggested that the shape of the Cl–H–Cl ion was very sensitive to its environment in the solid phase, being symmetrical like the bifluoride ion F–H–F in, for example, the tetraethylammonium salt but asymmetrical in the tetramethylammonium salt and in the cesium salt, although the critical band in the region of 200 cm^{-1}, assigned to the symmetric stretching mode of the Cl–H–Cl ion, was not within the range of frequency which was explored for the cesium salt. This conclusion becomes much more firmly based following studies of $CsHCl_2$ and $CsDCl_2$ by Stirling, Ludman and Waddington[11] using neutron spectroscopy. *Figure 10.2* compares their neutron spectra for the two salts, which were obtained using an incident beam of cold neutrons of wavelength 5.4 Å whose velocities after *gaining* energy by inelastic scattering were determined by time-of-flight analysis, observing the neutrons at a scattering angle of 82°. The important features in the spectra are the lines which are shown shaded and which are very much more intense in the hydrogen salt than for the deuterium version, from which we deduce that these particular lines involve motion of the hydrogen (or deuterium) atoms. The lines at 199 cm^{-1} and 195 cm^{-1} are the symmetrical stretching vibration and the fact that this involves motion of the hydrogen atom means that the HCl_2 ion *cannot* be symmetrical. It can then be established from further details of the Raman spectra that the ion is not only unsymmetrical but also bent. The intense peak at 675 cm^{-1} in the hydrogen spectrum is associated with a bending mode and it shifts in frequency in the deuterated salt, falling by

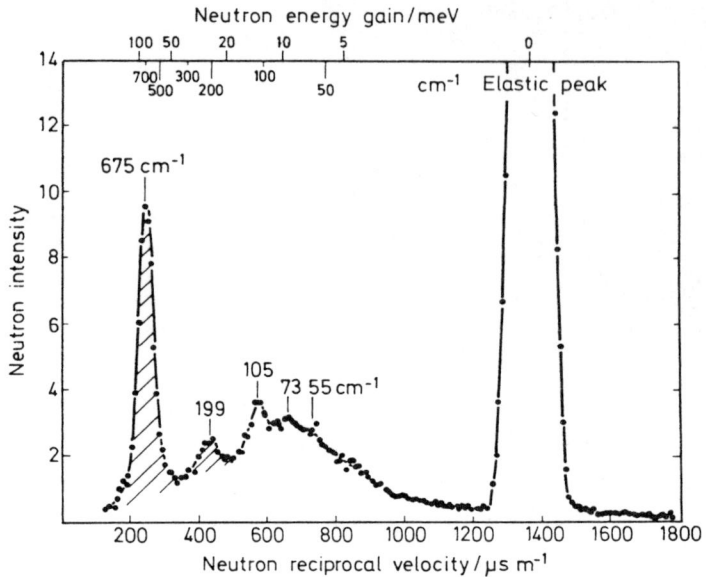
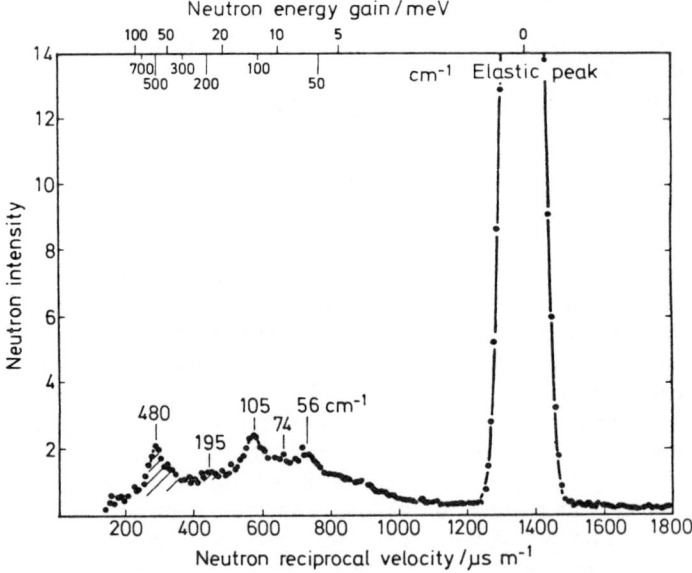

Figure 10.2 Energy-gain spectra for CsHCl$_2$ and CsDCl$_2$, using incident neutrons of wavelength 5.4 Å and a scattering angle of 74°. The spectral lines shown shaded are both much more intense for CsHCl$_2$ than for CsDCl$_2$, thus establishing that they involve H(D) motion (From Stirling, Ludman and Waddington[11], by courtesy of J. Chem. Phys.)

a ratio of about 1.4 as expected, as well as being much reduced in intensity. This peak for the bending mode is relatively weak in the IR and indeed its first harmonic at 1294 cm^{-1} was once assigned as the fundamental because of its strength. The antisymmetric stretching mode, which is observed in the IR at 1670 cm^{-1}, is at just too high a frequency to be observed in these neutron experiments.

It will have become evident that the experimental measurement of the neutron spectra can be made under two alternative conditions, for which the neutrons *gain* and *lose* energy, respectively. For a neutron energy-gain measurement it is usual to start with a beam of cold neutrons, and to investigate the increased energies of the scattered neutrons by time-of-flight analysis. Alternatively, neutron energy-loss can be employed by scattering a beam of monochromatic neutrons, choosing different energies successively, and assessing the slowed-down neutrons after scattering, with a beryllium-filter detector. It should be noted that the second method, whereby the neutrons *lose* energy, offers substantial advantages when the

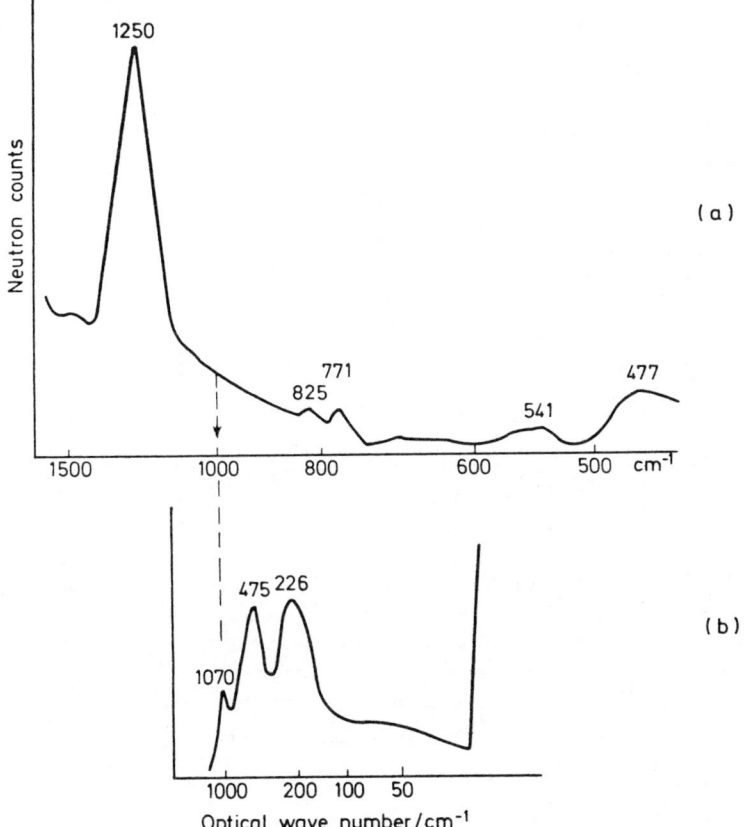

Figure 10.3 A comparison of (a) energy-loss (Ghosh, Waddington and Temme[12]*) and (b) energy-gain (Rush and Ferraro*[13]*) spectra for HCrO$_2$, indicating the improved resolution of the former technique for energy transfers greater than 400 cm*$^{-1}$

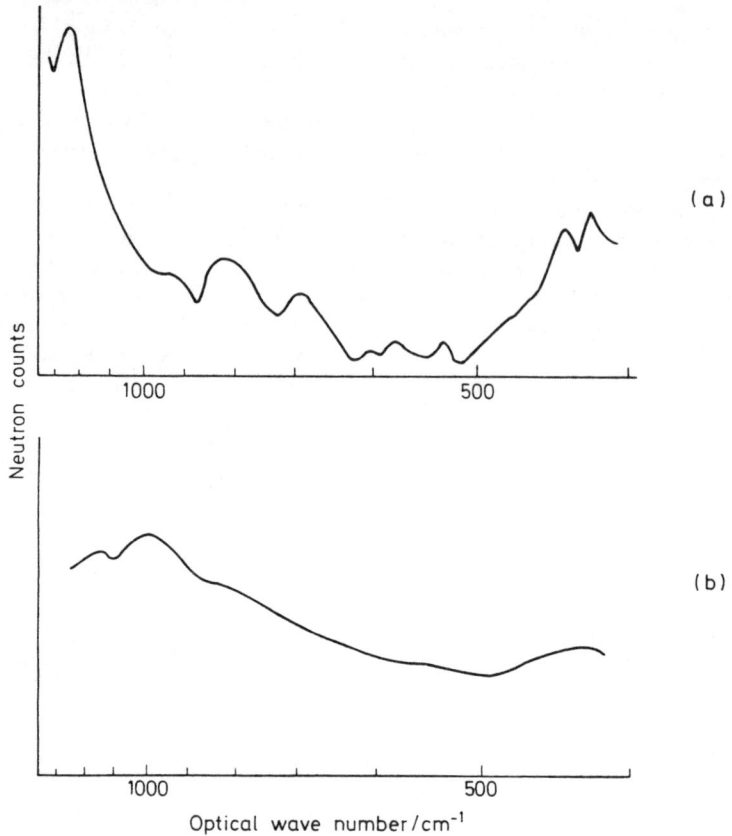

Figure 10.4 A comparison of energy-loss spectra of $KH(CO_2CCl_3)_2$ measured at (a) 80 K and (b) room temperature (After Ghosh, Waddington and Temme[12], by courtesy of I.A.E.A.)

vibration frequencies are greater than about 400 cm^{-1} (50 meV). This happens through two shortcomings of the energy-gain method. First, the resolution of the time-of-flight method is not very good at higher energies and, secondly, at ordinary temperatures there will be very low occupation of the higher-energy states from which the neutrons are required to draw energy. Quantitatively, this latter feature is expressed by the two different Boltzmann factors which are operative in equation 10.9 for gain and loss of neutron energy. Moreover, in the energy-loss technique a substantial improvement in the sharpness of the vibrational peaks in the spectrum can be achieved by cooling the scattering sample to the temperature of liquid nitrogen; with neutron energy-gain such cooling would of course impair even further the intensities for the vibrations of higher energy. As an example of the improvement in resolution at higher energies for the energy-loss technique, *Figure 10.3* contrasts the measured counting rates for $HCrO_2$ by energy-loss (Ghosh, Waddington and Temme[12]) and by energy-gain (Rush and Ferraro[13]). In the latter case there is very little

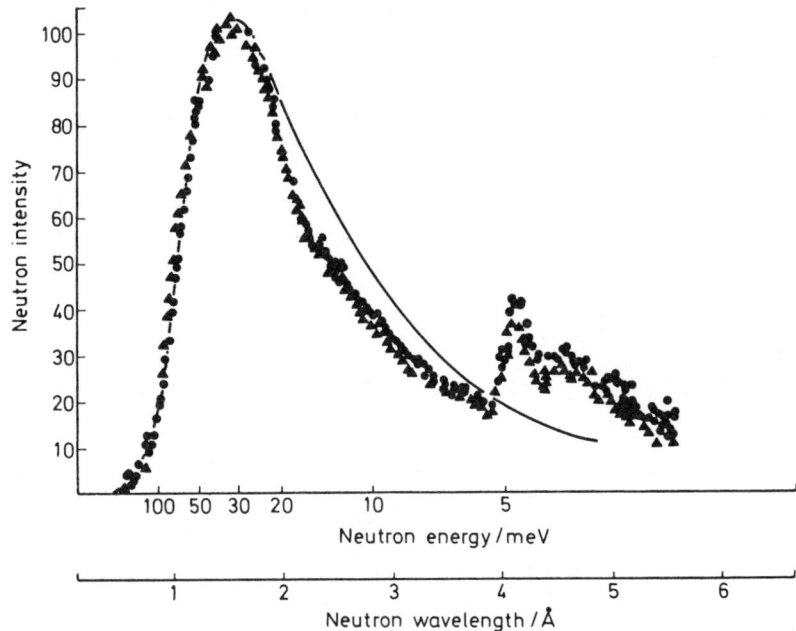

Figure 10.5 The energy-gain spectrum of NH_4ClO_4 at room temperature for incident neutrons of wavelength 4 Å and a scattering angle of 90°. The solid line is a theoretical curve for a free rotor with an effective mass of 2.1 (From Janik et al.[17], by courtesy of J. Phys. Chem. Solids)

discrimination available for vibrations above 1100 cm^{-1}, and the same conclusion is reached for the spectrum of $HCoO_2$ when it is examined by energy-gain (Delaplane et al.[14]). It is concluded (Temme and Waddington[15]) that the hydrogen bond in these materials is symmetric.

The substantial improvement in the resolution of the pattern when the sample is cooled to low temperature, using the energy-loss technique, is evident in *Figure 10.4*, which compares the spectra of $KH(CCl_3CO_2)_2$ obtained by Ghosh, Waddington and Temme[12] at room temperature and 80 K, respectively.

In general we may say that the energy-gain method, using cold neutrons and time-of-flight, is good for the range of energy transfer from 0 to 200 cm^{-1}, and most of the earlier neutron measurements were done in this way. Typical of these were many studies of the libration and rotation of ammonium and similar ions, such as the examination of $(NH_4)_2SO_4$ by Rush and Taylor[16], NH_4ClO_4 by Janik et al.[17] and PH_4I by Rush[18]. The latter two of these studies provide an interesting comparison. In the spectrum of ammonium perchlorate (*Figure 10.5*) there is a very broad peak near 240 cm^{-1} which is interpreted by assuming that the NH_4^+ ion is rotating freely, and the solid line in the figure is the theoretical curve for a free rotor with an effective mass of 2.1 m in terms of the theory of Krieger and Nelkin[19]. For phosphonium iodide (*Figure 10.6*), on the other hand, the very sharp peak at 335 cm^{-1} indicates that the PH_4^+

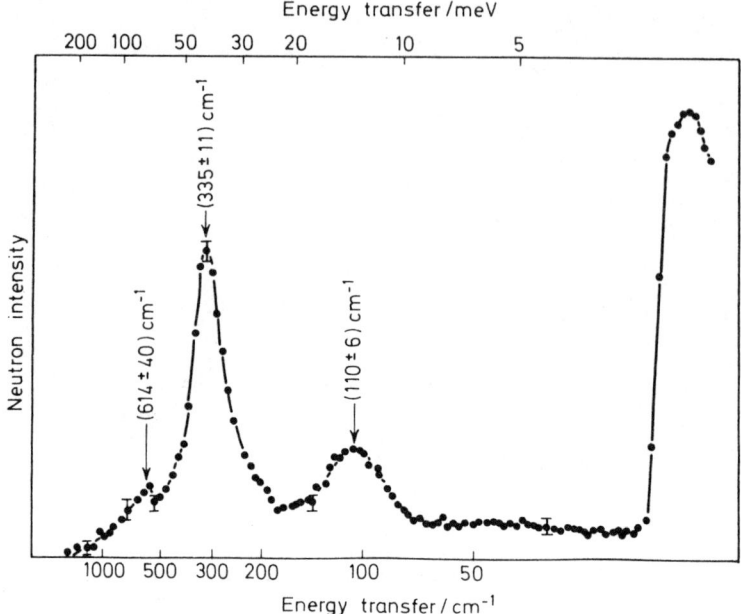

Figure 10.6 Energy-gain spectrum of PH_4I at room temperature for incident neutrons of wavelength 5 Å and a scattering angle of 60°. By contrast with Figure 10.5, the sharp peak at 335 cm^{-1} indicates that the PH_4^+ ion is tightly bound (From Rush[18], by courtesy of J. Chem. Phys.)

ion is very tightly bound. More recent measurements have increasingly used the energy-loss method, with a beryllium-filter detector, and can cover very well the range from 100 to 1200 cm^{-1}. The upper limit may be extended when experiments are carried out at high-flux reactors furnished with a hot source to increase the number of neutrons of short wavelength, possibly to 4000 cm^{-1}.

The salts K_2ReH_9 and Na_2ReH_9 provide an excellent example of materials for which the IR spectrum is not very informative, because of the selection rules which operate, but the neutron data of White and Wright[20] are quite conclusive. The neutron spectrum falls into two portions which are combined in *Figure 10.7* for the two salts. In the higher-frequency region of the Re—H bending modes, three frequencies can be distinguished, in contrast with the single band found by IR. In the range from 200 to 400 cm^{-1} four peaks exist for the potassium salt, two of them overlapping, but only two for the sodium salt. These are ascribed to libration of the whole ion and show that the motion is anisotropic with respect to the molecular axes and, moreover, that for K_2ReH_9 it is different for the two non-equivalent crystallographic sites in the structure, yielding four frequencies in all. For the sodium salt, which is not isomorphous, all the anionic sites are equivalent and only two vibration frequencies are found. The barriers to hindered rotation have been calculated from the observed neutron frequencies and are found to be 4.1, 7.2 and 8.8, 11.7 kcal mol^{-1} for the two sites in K_2ReH_9.

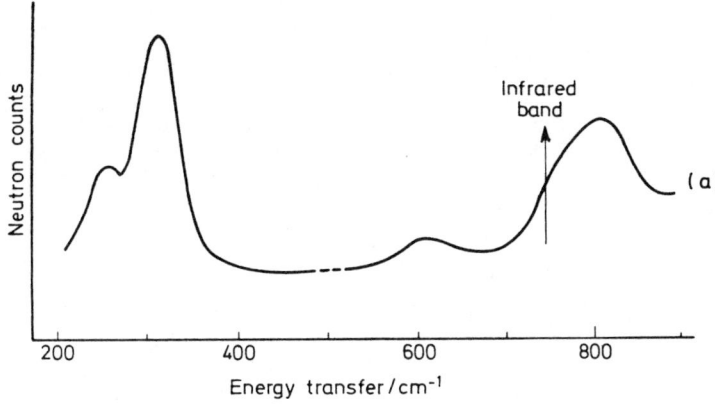
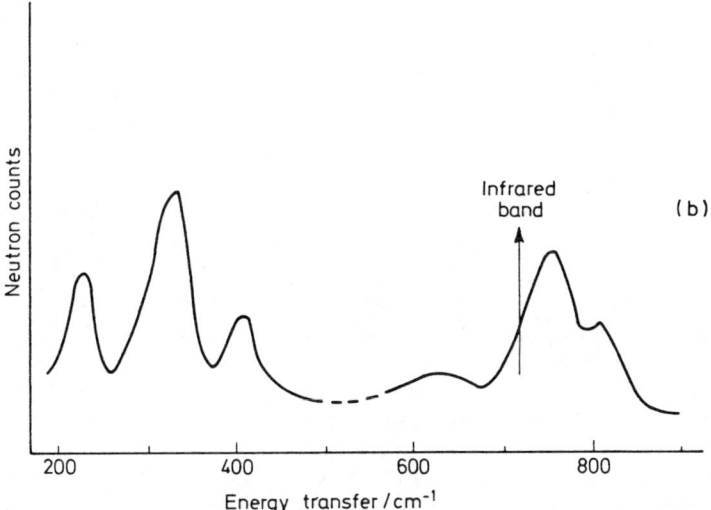

Figure 10.7 Comparison of the neutron energy-loss spectra of (a) Na_2ReH_9 and (b) K_2ReH_9. at 90 K using a beryllium-filter detector (After White and Wright[20], by courtesy of J. Chem. Soc. Faraday Trans. II)

10.3 Selective deuteration

In molecules which contain more than one type of hydrogen atom it is possible by selective deuteration to associate particular features of the spectrum with particular hydrogen atoms. This was first demonstrated for methanol, CH_3OH, by Aldred, Eden and White[21], whose results may be summarized by *Figure 10.8*. This figure shows spectra measured at an angle of scattering of 90° for the separate materials CH_3OH, CD_3OH and CH_3OD, in each case at room temperature. The important feature is the relative size of the broad peak centred on a time-of-flight of 500

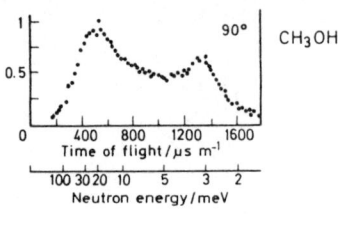
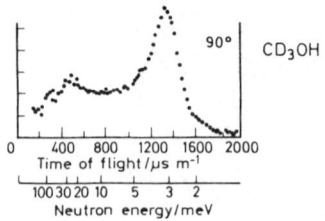
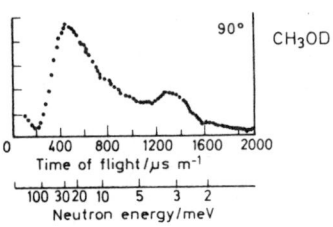

Figure 10.8 Energy-gain spectra for CH_3OH, CD_3OH and CH_3OD, for incident neutrons of wavelength 5.3 Å (0.003 eV) and time-of-flight analysis at a scattering angle of $90°$ (From Aldred et al.[21], by courtesy of Discuss. Faraday Soc.)

$\mu s\ m^{-1}$, which corresponds to an energy transfer of 160 cm^{-1}, and the sharper quasi-elastic peak at the right-hand side of the spectra. The broad peak falls substantially when the hydrogen atoms in the *methyl* group are replaced by deuterium, but there is not much change when deuterium is inserted in the *hydroxyl* group. It is therefore clear that the energy transfer of 160 cm^{-1} is associated with the methyl group. We shall give further examples of the use of isotopic substitution when we consider the scattering of neutrons by polymers, in a later chapter. It is also valuable to make comparisons of spectra when certain hydrogen atoms in a substance have been replaced by low-scattering atoms, such as fluorine, when this can be done without changing the crystal structure. In this way it has been shown that there is a broad peak, very similar to the peak for methanol, in the spectrum of acetic acid but not for CF_3CO_2H. Here too, then, the peak must be associated with hydrogen motion in the methyl group.

10.4 Surface chemistry and catalysis

A number of useful conclusions in the study of surface chemistry and catalysis have been reached by identifying the nature of adsorbed hydrogenous molecules. This has been done by examining their incoherent inelastic scattering for neutrons.

10.4.1 TRANSITION METAL COMPLEXES

A typical example is the elucidation of the role of the transition metals, particularly nickel and platinum, which catalyze hydrogenation reactions

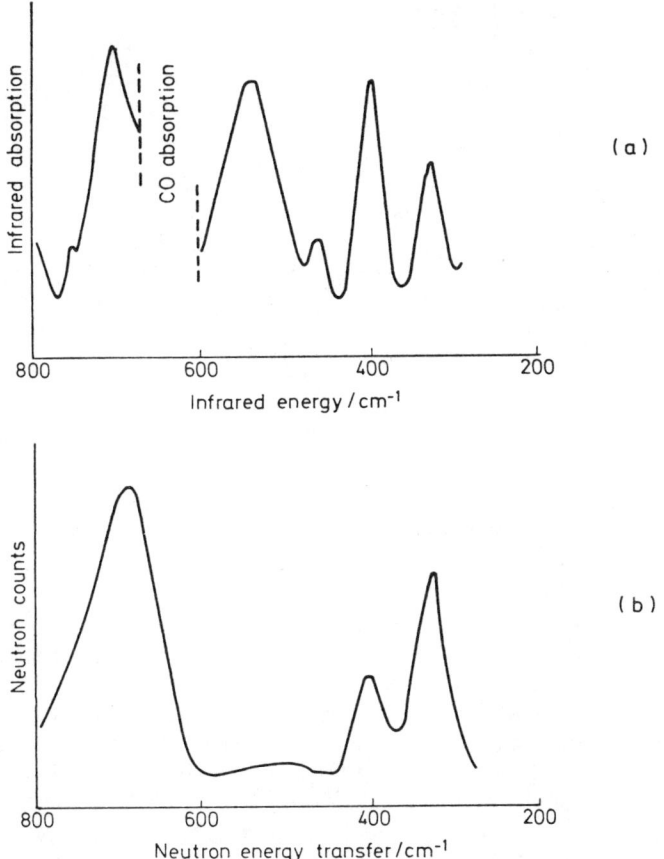

Figure 10.9 A comparison of (a) the infrared spectrum of HCo(CO)$_4$ with (b) the neutron energy-loss spectrum, measured for the solid at 90 K using a beryllium-filter detector. In particular the ratio of the bands at 700 cm^{-1} and 550 cm^{-1} is completely different for the two radiations. The former is a Co—H bending motion; the latter concerns Co—CO (From White and Wright[22], by courtesy of Chem. Commun.)

via the formation of carbonyl and other complex hydrides. One of the simplest of these is HCo(CO)$_4$ (cobalt carbonyl hydride), studied by White and Wright[22] and discussed further by White[23], which is important in industrial processes for producing aldehydes and alcohols. Comparison of the IR and neutron spectra in *Figure 10.9* underlines the different selection rules which operate for neutrons and can be used to suggest a structure for the HCo(CO)$_4$ molecule. The completely different ratio of the bands at 700 and 550 cm^{-1} for the two radiations indicates that these bands are concerned with Co—H and Co—CO motions, respectively.

In H$_2$Fe(CO)$_4$ (iron carbonyl dihydride) there is a reduction of molecular symmetry. The evidence for this is derived from *Figure 10.10*, which shows the splitting of the hydrogen bending vibration into in-phase and out-of-phase components, for the two hydrogen atoms, at 690 cm^{-1} and 760 cm^{-1}. At the same time a hydrogen bending mode for the

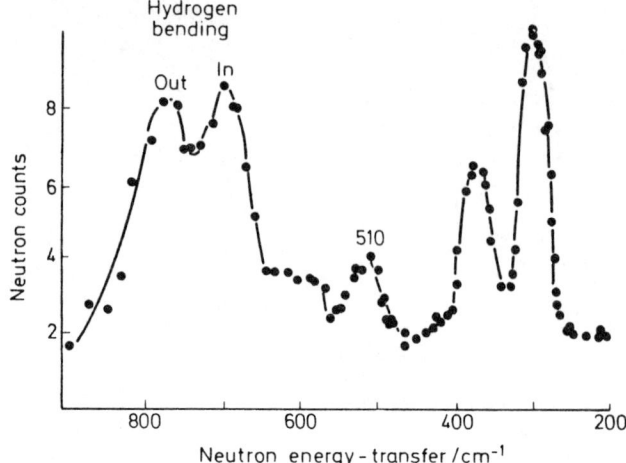

Figure 10.10 The energy-loss spectrum of $H_2Fe(CO)_4$ which, in comparison with that for $HCo(CO)_4$ in Figure 10.9, shows how the metal-hydrogen bending mode is split into in-phase and out-of-phase components, associated with the motion of the two hydrogen atoms (After White[23], by courtesy of I.A.E.A.)

reduced symmetry appears at 510 cm^{-1}; this did not appear in the neutron spectrum for $HCo(CO)_4$ but did appear in the IR spectrum of the latter at 468 cm^{-1}. The increases frequency for $H_2Fe(CO)_4$ is in keeping with the decrease in oxidation state of the metal atom.

As more complex examples of the compounds believed to participate in heterogeneous catalysis, $H_3Mn_3(CO)_{12}$ and $H_3Re_3(CO)_{12}$ have been examined (White and Wright[24], and White[23]). In these compounds the hydrogen atoms are believed to act as bridges between the manganese atoms and the carbonyl groups. Knowledge of the structure would be improved if detailed information were available regarding the vibrations of these hydrogen atoms. This is not forthcoming from IR or Raman spectra and the conventional procedure of identifying hydrogen vibrations by observing frequency shifts brought about by deuteration fails because of the overwhelming contributions to the spectrum from other motions. In the IR spectrum of $H_3Re_3(CO)_{12}$, *Figure 10.11*, the hydrogen contributions are in fact represented by the weak line A at 339 cm^{-1} and the shoulder B at 625 cm^{-1} and both of these features are lost in the spectrum of the deuterated compound. Indeed, their identification is only made possible by the neutron spectrum, shown in *Figure 10.11(b)* for $H_3Mn(CO)_{12}$, in which these hydrogen bending vibrations are the predominant features.

10.4.2 MOLECULAR-SIEVE ZEOLITES

The molecular-sieve zeolites form an important group of catalysts having uniform pores of molecular dimensions, which provide a very high surface area but are accessible only to molecules up to a certain size. The

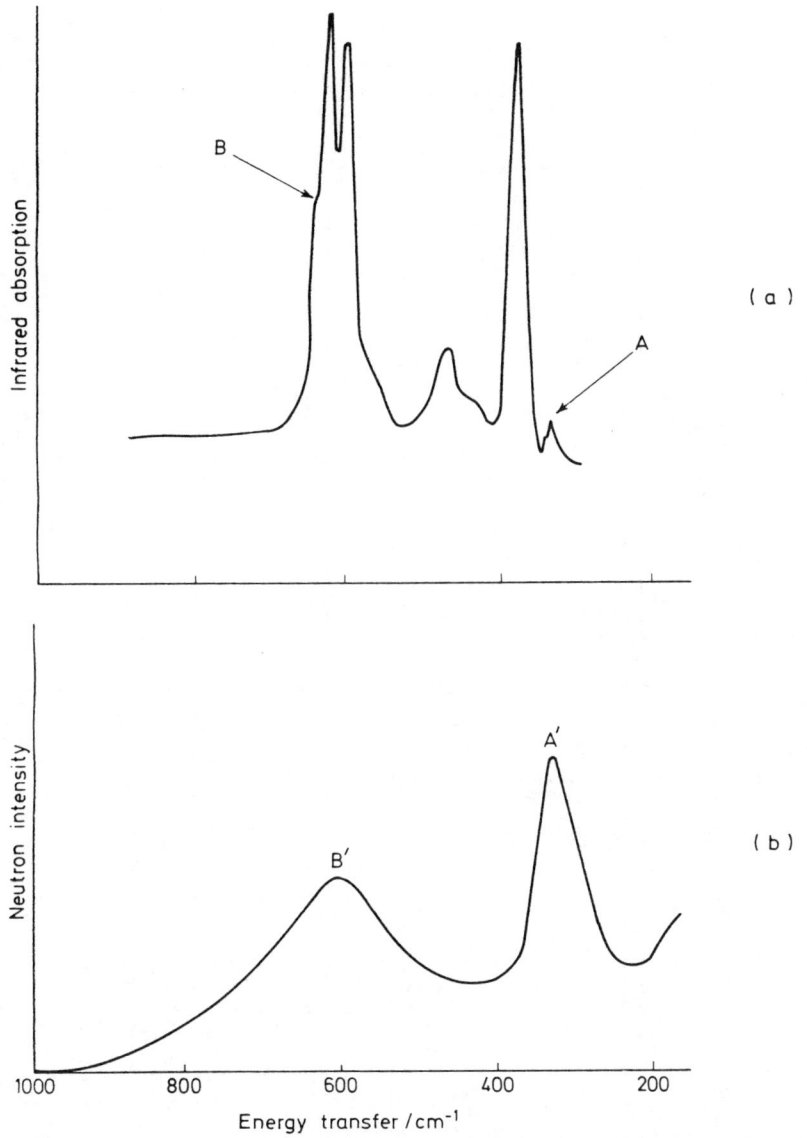

Figure 10.11 Comparison of (a) the infrared spectrum of $H_3Re_3(CO)_{12}$ with (b) a neutron spectrum for $H_3Mn_3(CO)_{12}$. In the former the hydrogen contributions are limited to the small peak at A and the shoulder B on the main band. In the neutron spectrum these small details become the predominant features (From White and Wright[24], by courtesy of Chem. Commun.)

adsorption of hydrogenous species by them can be studied very well by neutron inelastic scattering, particularly in view of the very limited scattering produced by the three-dimensional network of silica and alumina tetrahedra. We can contrast the behaviour of methane and water when they are adsorbed in the cavities. *Figure 10.12* (White[23]) shows energy analyses

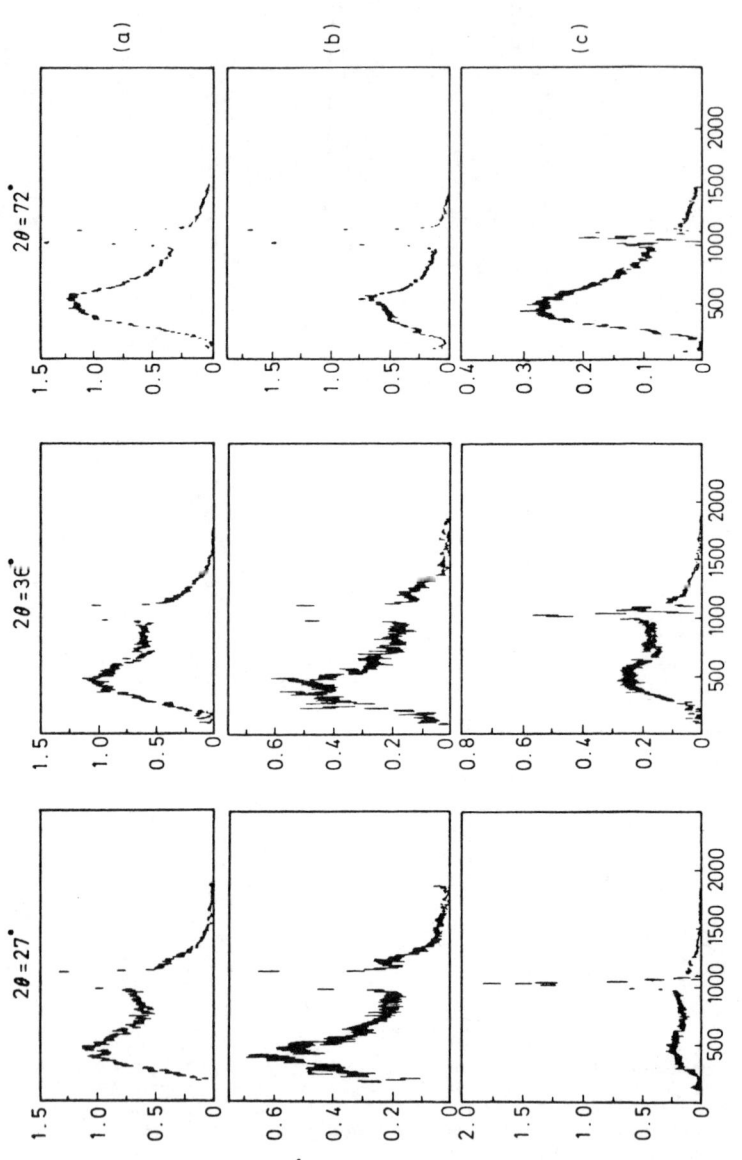

Figure 10.12 Neutron energy-gain spectra illustrating the absorption behaviour of a lanthanum-exchanged sodium Y zeolite for methane. The three columns of spectra refer to scattering at angles of 27, 36 and 72°, respectively. In each case the upper row (a) shows results for zeolite, container and CH_4. The middle row (b) is for the outgassed zeolite in the container and the bottom row (c), which is a difference plot, is ascribed to the methane alone (From White[23], by courtesy of I.A.E.A.)

of the neutrons which are scattered at three different angles for (*a*) a zeolite with methane, (*b*) an outgassed zeolite and (*c*), by difference, the scattering due to the methane alone. The pattern at (*c*) for the scattering at $2\theta = 72°$ is that predicted by the Kreiger-Nelkin[19] theory for freely rotating CH_4 groups. Likewise it can be compared with the patterns from freely rotating NH_4 groups, deduced by Janik *et al.*[17] in ammonium perchlorate. At smaller angles of scattering the broad peak which is centred on the elastic peak near to a flight time of 1000 μs m^{-1} is accounted for in the same way. Thus most of the methane molecules are in the gaseous phase.

On the other hand, when the cavities in the zeolites are partially filled with water the spectrum is quite different. As *Figure 10.13* shows, the scattering is not dissimilar to that of liquid water and there is a quasi-elastic peak A from which the diffusion coefficient of the water molecules can be deduced. The typical water peak B, for an energy transfer of 480 cm^{-1} from the water molecules, is present in the zeolite; on the other hand, the peaks in the region 10–60 cm^{-1} are substantially reduced for the water in the cavities compared with normal water.

10.4.3 WATER LAYERS IN CLAY MINERALS

A somewhat related topic is the study of the water layers which can exist between the silicate sheets in certain clay minerals such as vermiculite and montmorillonite: water layers of controllable thickness from zero up to about 100 Å can be established and it is likely that the structure of the silicate confers some degree of order on the water structure. The neutron inelastic scattering will largely be determined by the vibrational modes which involve hydrogen atoms and by the diffusive motion of the water molecules within the layers. *Figure 10.14* shows[23] that the inelastic-neutron-scattering spectra depend on the thickness of the water layers; a spectrum for water itself is included for comparison. The broad peak at 48 cm^{-1} in water is much attenuated in those layers which are only one or two molecules thick, in a similar way to that shown in *Figure 10.13* for the zeolites. This suggests that in the zeolites the surfaces are only covered by a monomolecular, or possibly a bimolecular, film of water. It will also be seen from *Figure 10.14* that the quasi-elastic peak broadens as the thickness of the water layer increases. This broadening ΔE is a consequence of the energy exchange introduced by the Doppler effect and it can be shown that

$$\Delta E = 2\hbar D \left(\frac{4\pi}{\lambda} \sin \theta \right)^2 = 2\hbar D Q^2 \qquad (10.11)$$

where λ is the mean wavelength and D is the diffusion coefficient. Measurements of patterns such as those in *Figure 10.14* shows that log D is proportional to the reciprocal of the thickness of the water layer. This is illustrated in *Figure 10.15* for a number of different clay minerals and

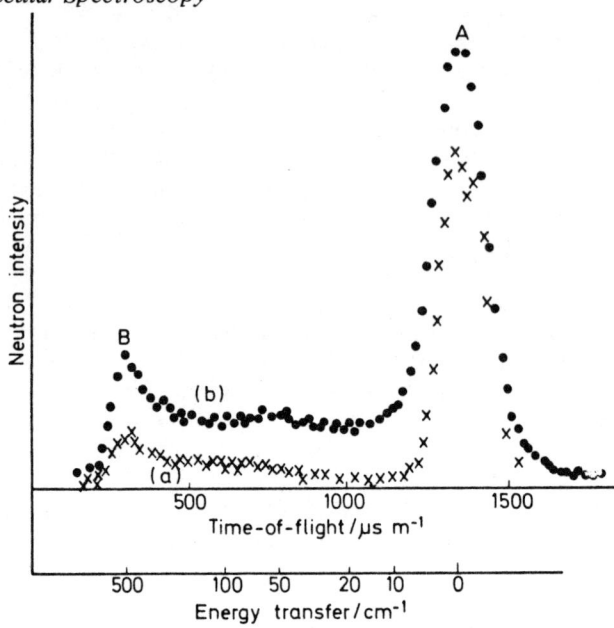

Figure 10.13 Neutron time-of-flight spectra for the scattering of neutrons of wavelength 5.3 Å by (a) for points x, a zeolite partly filled with water and (b), for points •, water itself (From White[25], by courtesy of I.A.E.A.)

it is found that the line shown there extrapolates, as $1/d$ approaches zero, close to the macroscopic value for D, which equals 2.3×10^{-5} cm^2 s^{-1}. The linear relationship is consistent with the suggestion of Olejnik and White[25] that the water meniscus around the circumference of the clay platelets is concave outwards, with a contact angle of zero and a radius of $d/2$. The linearity also makes it unlikely that there is any marked long-range order of the water molecules close to the silicate surface.

10.4.4 IONIC SOLUTIONS

The determination of the diffusion coefficient from the width of the quasi-elastic scattering can be used to compare the structures of pure water and ionic solutions. In an ionic solution we have to consider the diffusion of both the solvent molecules and the solute ions. White[24] has shown that the proton diffusion, which is what the neutrons measure, is greater in solutions of lithium chloride than it is in pure water. On the other hand, in solutions of potassium iodide the motion is less than in pure water. In each case the neutron determinations, which relate to a time of observation of about 10^{-12} s, agree with macroscopic measurements such as tracer diffusion. In order to study the motion of the solute ions it is necessary for the neutron scattering by these to be very much greater than that by the solvent molecules. This can be achieved by using heavy water, which has very small incoherent scattering, as the

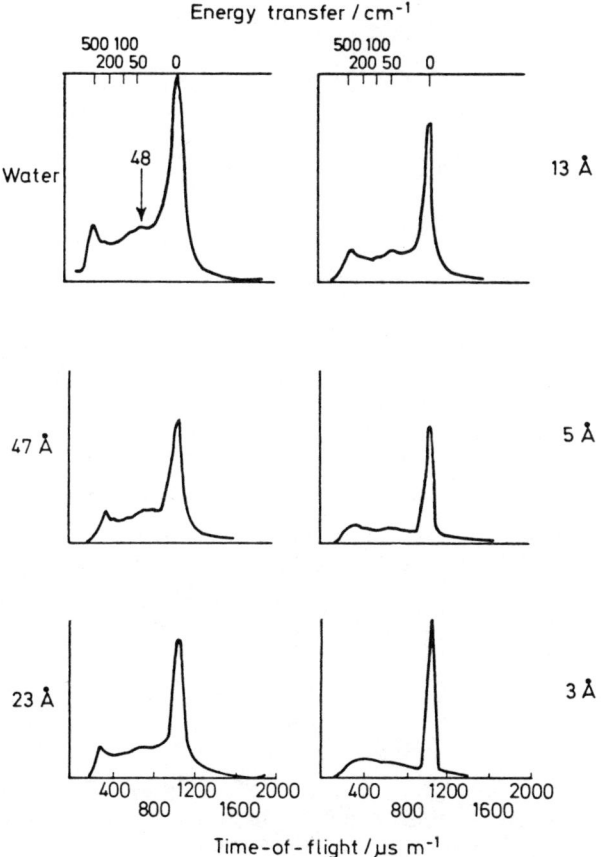

Figure 10.14 Neutron time-of-flight spectra for incident neutrons of wavelength 4.2 Å scattered at an angle of 90° from pure water, and from water layers of various thicknesses between the aluminosilicate sheets of lithium montmorillonite. The scale at the top of the diagram gives the energy gain of the neutrons in cm⁻¹ (From White[23], by courtesy of I.A.E.A.)

solvent and tetramethylammonium bromide as the solute. The $N(CH_3)_4^+$ ion then has the enormous incoherent scattering which comes from its hydrogen atoms. It is found that the diffusion of the solute atoms is significantly lower than that for the solvent protons but the value obtained with neutrons is about three times as great as from nuclear magnetic resonance data. This has been explained as an effect due to rotational diffusion, either of the ion as a whole or within the methyl groups. Such rotations will be 'noticed' by the neutron, on its timescale of 10^{-12} s, but not on the longer timescale of other techniques which record the motion of the centre of mass of the ion over distances large compared with its radius.

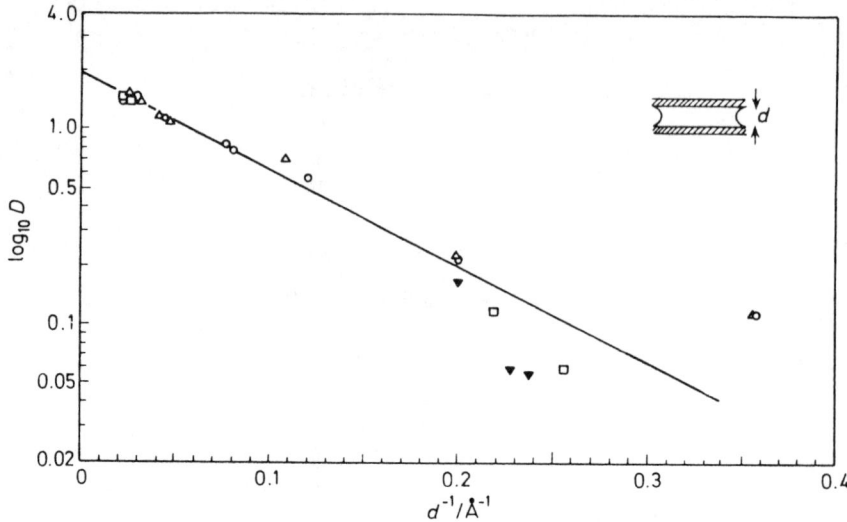

Figure 10.15 The dependence of the self-diffusion constant D of water in clay minerals, as determined from the width of the quasi-elastic peak in the neutron spectrum, on the thickness d of the water layer. The experimental points are indicated by ○ lithium montmorillonite, □ lithium vermiculite, △ sodium montmorillonite and ▼ sodium (From Olejnik and White[25], by courtesy of Nature)

References

1. VAN HOVE, L., *Phys. Rev.*, **93**, 268; **95**, 249; and **95**, 1374 (1954)
2. MARSHALL, W. and LOVESEY, S.W., *Theory of Thermal Neutron Scattering*, Clarendon Press, Oxford (1971)
3. DOLLING, G. and POWELL, B.M., *Proc. Roy. Soc.*, **A319**, 209 (1970)
4. REYNOLDS, P.A., KJEMS, J.K. and WHITE, J.W., in *Neutron Inelastic Scattering*, 195, I.A.E.A., Vienna (1972)
5. LOMER, W.M. and LOW, G.G.E., in *Thermal Neutron Scattering*, ed. P.A. Egelstaff, 1, Academic Press, London (1965)
6. WINDSOR, C.G., in *Chemical Applications of Thermal Neutron Scattering*, ed. B.T.M. Willis, 1, Clarendon Press, Oxford (1973)
7. WHITE, J.W., in *Chemical Applications of Thermal Neutron Scattering*, ed. B.T.M. Willis, 49, Clarendon Press, Oxford (1973)
8. BOUTIN, H., SAFFORD, G.J. and BRAJOVIC, V., *J. Chem. Phys.*, **39**, 3135 (1963)
9. COLLINS, M.F., HAYWOOD, B.C. and STIRLING, G.C., *J. Chem. Phys.*, **52**, 1828 (1970)
10. EVANS, J.C. and LO, G.Y-S., *J. Phys. Chem.*, **70**, 11 (1966)
11. STIRLING, G.C., LUDMAN, C.J. and WADDINGTON, T.C., *J. Chem. Phys.*, **52**, 2730 (1970)
12. GHOSH, R.E., WADDINGTON, T.C. and TEMME, F.P., in *Neutron Inelastic Scattering*, 345, I.A.E.A., Vienna (1972)
13. RUSH, J.J. and FERRARO, J.R., *J. Chem. Phys.*, **44**, 2496 (1966)

14. DELAPLANE, R.G., IBERS, J.A., FERRARO, J.R. and RUSH, J.J., *J. Chem. Phys.*, **50**, 1920 (1969)
15. TEMME, F.P. and WADDINGTON, T.C., *J. Chem. Phys.*, **59**, 817 (1973)
16. RUSH, J.J. and TAYLOR, T.I., in *Inelastic Scattering of Neutrons*, Vol. II, 333, I.A.E.A., Vienna (1965)
17. JANIK, J.A., JANIK, J.M., MELLOR, J. and PALEVSKY, H., *J. Phys. Chem. Solids*, **25**, 1091 (1964)
18. RUSH, J.J., *J. Chem. Phys.*, **44**, 1722 (1966)
19. KRIEGER, T.J. and NELKIN, M.S., *Phys. Rev.*, **106**, 290 (1957)
20. WHITE, J.W. and WRIGHT, C.J., *J. Chem. Soc. Faraday Trans. II*, **68**, 1414 (1972)
21. ALDRED, B.K., EDEN, R.C. and WHITE, J.W., *Discuss. Faraday Soc.*, **43**, 169 (1967)
22. WHITE, J.W. and WRIGHT, C.J., *Chem. Commun.*, 970 (1970)
23. WHITE, J.W. in *Neutron Inelastic Scattering*, 315, I.A.E.A., Vienna (1972)
24. WHITE, J.W. and WRIGHT, C.J., *Chem. Commun.*, 971 (1970)
25. OLEJNIK, S. and WHITE, J.W., *Nature Phys. Sci.*, **236**, 15 (1972)
26. WHITE, J.W., *Ber. Bunsengesellschaft Phys. Chem.*, **75**, 379 (1971)
27. EGELSTAFF, P.A., DOWNES, J.S. and WHITE, J.W., *Chem. and Ind.*, 306, (1967)

11

POLYMERS

Both the opportunities and limitations of neutron scattering methods are illustrated by their application to the study of polymers. Essentially, polymers consist of long chains of repeating subunits such as $-CH_2-CH_2-$, as in polyethylene. These chains may possess side chains of variable length and may be linked together to give three-dimensional networks. In practice a chain generally includes from 100 to 100 000 units. The forces within the chains are greater by an order of magnitude than the forces between neighbouring chains and, indeed, many elementary aspects of the structural behaviour of polymers can be approximated by a theoretical model which assumes infinite isolated chains. Polymeric materials can exist in the three states of rubber, glass and crystalline, representing an increasing degree of order. In the glassy state, obtained by chilling the rubber, the main chain of the polymer is frozen in position, albeit in an irregular fashion, but side groups may still be in motion. In the 'crystalline' state the crystallites may account for 40–95% of the material, being in general randomly oriented in a matrix of amorphous material. In polymers such as orthorhombic polyethylene, $(-CH_2-CH_2-)_n$, and hexagonal polyfluorethylene, $(-CF_2-CF_2-)_n$, a high degree of alignment of the c axes of the crystallites can be produced by drawing and rolling. By various combinations of these processes it is possible to produce partial orientation of the a and b axes also, but their alignment is not likely to be better than 15°. It is only partially possible therefore to study the full details of the phonon spectrum, including its dependence on the phonon direction, as can be done for perfect ionic or metal crystals by observing the *coherent* inelastic scattering of neutrons. It will be possible, however, in an examination of *incoherent* inelastic scattering, to observe the presence and contribution of particular vibrations. In each case the virtue of neutron spectroscopy, in contrast to IR and Raman techniques, will arise from the absence of the usual selection rules and from the diagnostic power of comparing data for normal and deuterated materials.

It will be convenient to review some of the recent neutron work, aiming to understand the dynamics of polymer structures, in terms of *Figure 11.1,* which shows the results of the calculations of Tasumi and Shimanouchi[1] for infinitely long chains. For isolated chains there are two acoustical branches: ν_9 is a torsional vibration around the C–C bonds, reaching up to about 200 cm^{-1}, and ν_5 is a skeletal deformation or 'accordion' motion of the chain with a limit of about 500 cm^{-1}. The

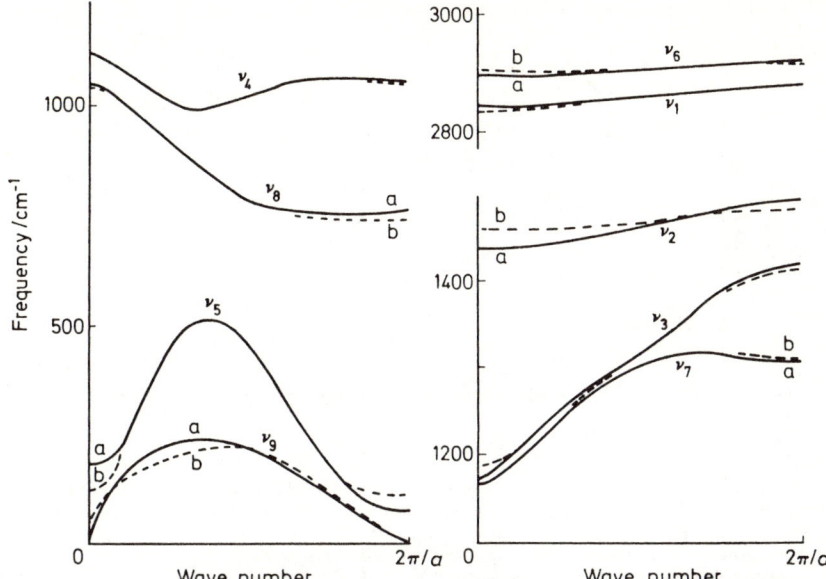

Figure 11.1 Calculated dispersion curves for polyethylene. The splitting of the modes into the components a *and* b *is the result of intermolecular forces between the infinitely long chains (From Tasumi and Shimanouchi, by courtesy of J. Chem. Phys.)*

high-frequency optical branches are relatively flat. When allowance is made for intermolecular forces between neighbouring chains, each mode is split into two components a and b. The changes are most drastic for ν_5 and ν_9 and particularly in the low-frequency region: for these modes the new curves are indicated in *Figure 11.1*. Both the skeletal deformation and the torsional vibration are mixed in ν_5 and ν_9, although these do retain their predominant characters: $\nu_9{}^a$ remains zero when q equals zero or $2\pi/a$, where a is the period along the chain, but $\nu_5{}^a$, $\nu_5{}^b$ and $\nu_9{}^b$ no longer go to zero and their intercepts on the frequency axes correspond to interchain vibrations. When the density of states function is computed from the observed spectra it is found that both the maxima in the dispersion curves for ν_5 and ν_9 and their intercepts at $q = 0$ and $q = 2\pi/a$ correspond to maxima in the frequency distribution.

Following earlier work by Boutin[2] and others, mainly with polycrystalline material, Myers, Summerfeld and King[3] clearly identified the ν_9 and ν_5 modes for polyethylene in measurements with a triple-axis spectrometer using material which was 86% crystalline and had been oriented by stretching to 9.5 times its original length. The results are illustrated in *Figure 11.2*, which shows two frequency distribution curves for which the momentum transfer directions Q are, respectively, parallel and perpendicular to the direction of the polymer chains in the sample. The peak at 525 cm^{-1} occurs only when Q is along the polymer chains, thus identifying it as ν_5 the accordion vibration. The peak at 195 cm^{-1}

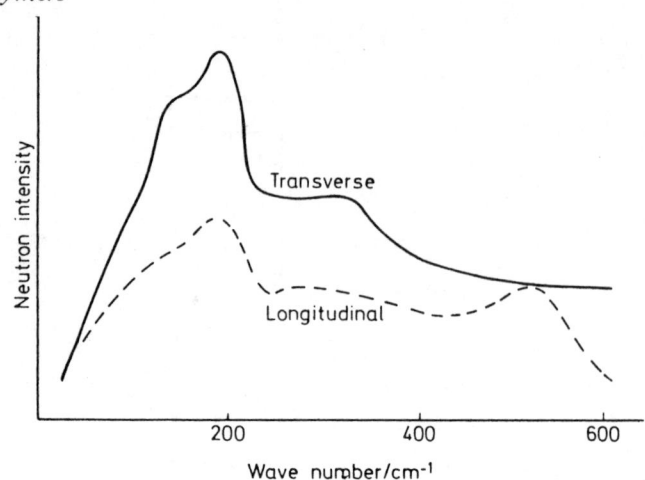

Figure 11.2 Energy-loss neutron spectra for stretched and oriented polyethylene. Results are shown for both transverse and longitudinal vibrations. The peak at 195 cm^{-1} is ν_9 the torsional vibration; that at 525 cm^{-1} is the longitudinal stretching mode which would be expected to have no transverse component (After Myers, Summerfield and King[3], by courtesy of J. Chem. Phys.)

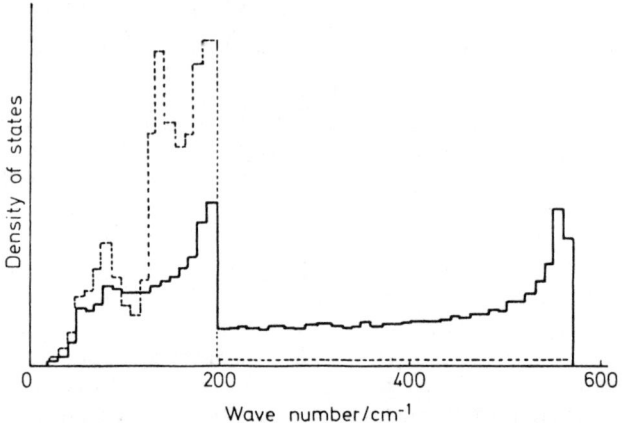

Figure 11.3 A histogram showing the calculated density-of-states function, $Z(\omega)$, for oriented polyethylene. The solid line is for momentum transfer longitudinally, parallel to the chain axis; the dotted line is for momentum transfer perpendicular to the chain axis (From Kitagawa and Miyazawa[4], by courtesy of J. Chem. Phys.)

is found in both measurements, but it is reduced to 56% when Q is at right-angles to the polymer chains and was identified as ν_9, the torsional vibration. *Figure 11.2* makes evident the anisotropy in the inelastic neutron scattering cross-section, which is dependent on the direction of the momentum transfer. From the curves in *Figure 11.1* as modified to take account of the intermolecular forces, it is possible to compute a curve for the density of states, to be correlated with the incoherent neutron spectra. *Figure 11.3* (Kitagawa and Miyazawa[4]) shows this in the form of a histogram for oriented polyethylene, for the two directions of momentum transfer, parallel and perpendicular respectively to the c axis. In good agreement with the experimental results, the longitudinal peak at 195 cm^{-1} is only half as high as the corresponding transverse peak, whereas the peak near 560 cm^{-1} is almost completely longitudinal.

From time-of-flight measurements, Twistleton and White[5] have provided full data in the region below 250 cm^{-1} for the incoherent scattering by

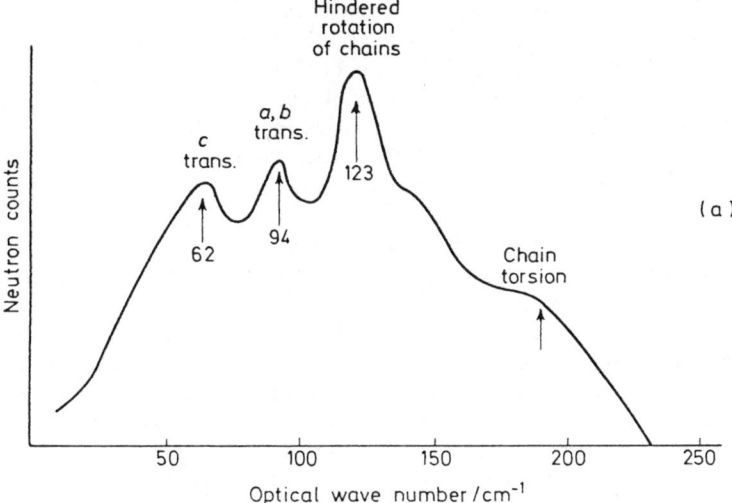

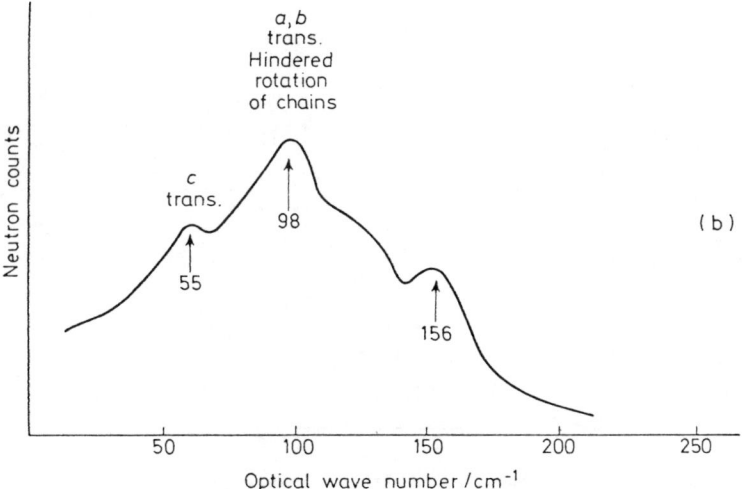

Figure 11.4 Comparison of the neutron energy-gain spectra for (a) normal and (b) deuterated polyethylene at 77 K in the region of low energy transfers (From Twistleton and White[5], by courtesy of I.A.E.A.)

both normal and deuterated polyethylene and have identified four peaks which are in good agreement with calculations. These spectra measured at 77 K are shown in *Figure 11.4*. The mode of lowest frequency, at 62 cm^{-1} and 55 cm^{-1} for hydrogen and deuterium, respectively, is due to antiparallel movement of the chains along the c axis; that at 94 cm^{-1} and 98 cm^{-1} is the mean result of antiparallel vibrations along the a and b axes. At 123 cm^{-1} there is a peak in the hydrogen spectrum due to hindered rotation of the chains against each other, but in the deuterated spectrum this is at a substantially lower frequency and overlaps with the translation mode at 98 cm^{-1}. Peaks at 190 cm^{-1} and 156 cm^{-1}, for hydrogen and deuterium, are due to torsional motion of the chains. There is a discrepancy in the value of 123 cm^{-1} observed for the hindered rotatory motion: the calculations of Shimanouchi predict a peak at 160 cm^{-1} which is well beyond the experimental error. This suggests that the neutron data may contribute to an improved picture of the intermolecular forces.

Direct determination of the dispersion curves rests on observation of the *coherent* scattering from highly oriented material and has been achieved by using deuterated samples. Feldkamp et al.[6] studied highly stretched samples of deuterated polyethylene for which all the c axes were oriented within an angle of 9°. The results are only unambiguous for phonons propagating along the c axis, since there was little orientation of the a or b axes. The measurements were in substantial agreement with the dispersion curves calculated theoretically for an isolated chain, confirming that the forces between adjacent chains are much weaker than the intramolecular forces. Fuller data have been obtained by Twistleton

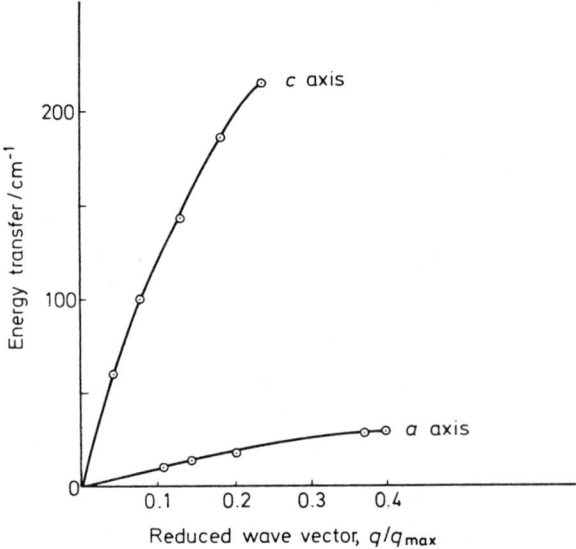

Figure 11.5 A comparison of the dispersion curves for Teflon for the c and a directions obtained respectively by LaGarde, Prask and Trevino[8] and Twistleton and White[7] from the coherent scattering of oriented fibres

and White[5], not only for polyethylene but also for Teflon, the substituted material polytetrafluorethylene (Twistleton and White[7]) which is hexagonal with a helical chain having 15 repeat units in 7 turns. As an example we can consider the results for Teflon which, with a composition $(-CF_2-)_n$ not containing hydrogen, gives an inelastic scattering spectrum which is mainly coherent. *Figure 11.5* compares the dispersion curves deduced for lattice vibrations perpendicular to the chain axis by Twistleton and White[7] and for vibrations along the chain by La Garde, Prask and Trevino[8], using oriented extruded fibres. In each case the slope of the curve in the linear region near the axis gives the velocity of sound, and hence the elastic constant, in the appropriate direction. For the vibrations

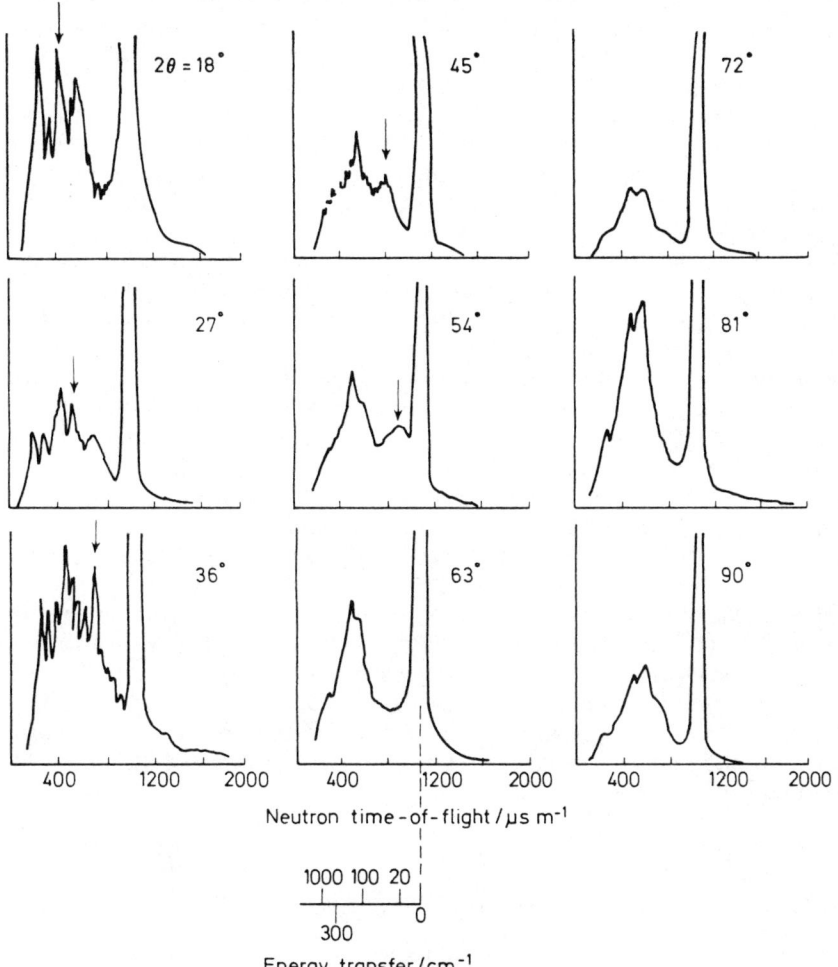

Figure 11.6 Neutron energy-gain spectra, using incident neutrons of wavelength 4.2 Å, for polycrystalline polydeuteroethylene, illustrating the dependence of the coherent scattering on the value of the scattering angle (From Holliday and White[9], by courtesy of J. Pure Appl. Chem.)

along the chain axis the velocity is 9.74×10^5 cm s^{-1} and C_{33} is 222×10^{10} dyn cm^{-2}, whereas at right-angles to the axis the velocity is 2.77×10^5 cm s^{-1} and C_{11} is 18.2×10^{10} dyn cm^{-2}. Thus we have a very direct demonstration of the anisotropy of the elastic constant, with a ratio of about 12 : 1 for C_{33}/C_{11}. For deuterated polyethylene the coherent neutron data give a much larger ratio of about 20 : 1.

It is to be noted that the interpretation of the *coherent* inelastic scattering spectra from partially polycrystalline material, as distinct from single-crystal samples of metals and simple ionic compounds, is complicated by the dependence of the scattered spectra on the momentum transfer in the scattering process. This is well illustrated in the observations of the coherent scattering of deuterated polyethylene at different angles of scattering described by Holliday and White[9]. *Figure 11.6* shows spectra recorded at nine different angles of scattering and it will be noticed that the sharp peak near 500 μs m^{-1} for $2\theta = 18°$ moves to smaller values of energy transfer as the scattering angle increases: when $2\theta = 54°$ the peak is near 1000 μs m^{-1} and by $2\theta = 63°$ it has merged with the elastic peak. This behaviour is characteristic of the dispersion relations associated with the phonons connected with the lattice vibrations. On the other hand, peaks due, for example, to the torsional vibrations of the chain remain largely unchanged in frequency at all angles of scattering.

In the glassy state of a polymer, which is produced by chilling a rubber, the main chain of the polymer molecule is frozen in some random configuration but side groups may still be in motion, and this motion is important in determining some of the mechanical and dielectric properties of the materials. This motion may consist predominantly of a torsional vibration of a CH$_3$ group and this will be very readily detected with neutrons because of the large amplitude of motion and the large incoherent scattering of the protons. From the observed frequencies of vibration it is possible to calculate the value of the potential barrier which hinders complete internal rotation of the methyl groups. These torsional modes are, in principle, active in IR and Raman spectra but the changes in dipole moment and polarizability are so small that the corresponding bands are very weak and difficult to identify. Moreover, in the neutron measurements the identification of the vibrations can be confirmed by observing the changes in the spectra when the CH$_3$ groups are replaced by CD$_3$. This is indicated very clearly in a study of Perspex, poly(methyl methacrylate), and some of its derivatives by Higgins, Allen and Brier[10]. Measurements were made with

$$\begin{array}{ccc} & \text{CH}_3 & \text{CH}_3 & \text{CH}_2\text{Cl} \\ & | & | & | \\ -\text{CH}_2-\text{C}- & , \quad -\text{CH}_2-\text{C}- \quad \text{and} \quad -\text{CH}_2-\text{C} \\ & | & | & | \\ & \text{CO}_2\text{CH}_3 & \text{CO}_2\text{CD}_3 & \text{CO}_2\text{CH}_3 \end{array}$$

I II III

poly(methyl methacrylate)   poly(α-chloromethyl methacrylate)

Duplicate measurements were made with each of these materials when swollen with deuterated chloroform, which performs the function of separating the polymer molecules by what, to the neutrons, are practically invisible molecules of $CDCl_3$. The effect of this treatment on material I is indicated in *Figure 11.7.* The spectra of the swollen materials are substantially sharpened and the region between 100 and 400 cm^{-1} is seen to be resolved into three bands centred at 100, 240 and 350 cm^{-1}. The nature of the first and third of these bands is revealed in *Figure 11.8*, which compares the spectra of the swollen versions of the three materials whose formulae are shown above. The band at 100 cm^{-1} appears for materials I and III but not for II and is accordingly associated with the ester methyl group in the CO_2CH_3 side chain. On the other hand, the band at 350 cm^{-1} occurs for I and II but not for III; hence it is due to the α-methyl side group. The fact that these spectral frequencies are not changed when the materials are swollen suggests that they are intramolecular in origin and they are assigned to the torsional vibration of the methyl groups. The values of the potential function for the hindered rotation of the groups can be calculated from these frequencies.

In all these measurements, observations of neutron intensity were made at several different angles of scattering 2θ, ranging from $18°$ to $90°$, and

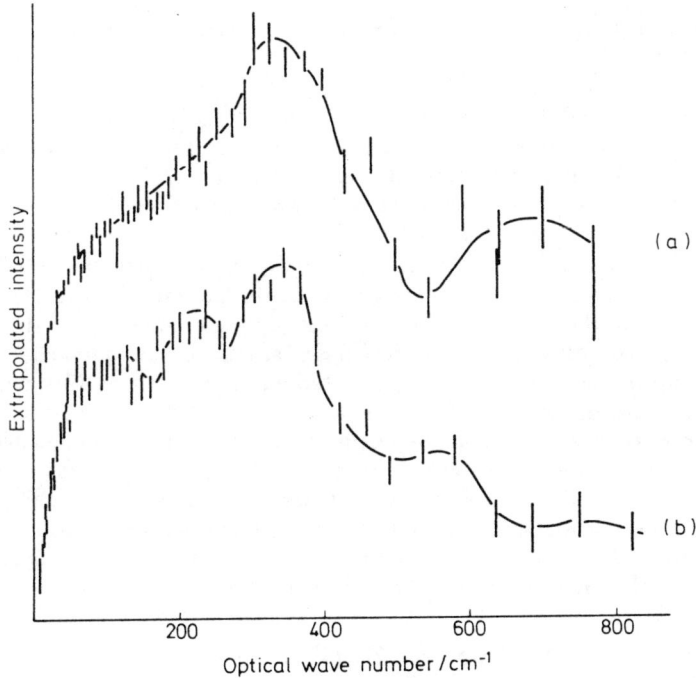

Figure 11.7 Curves of extrapolated neutron intensity for the energy gain spectra, using incident neutrons of wavelength 4.2 Å, for (a) perspex and (b) perspex swollen with deuterated chloroform, $CDCl_3$ (From Higgins, Allen and Brier[10], by courtesy of Polymer)

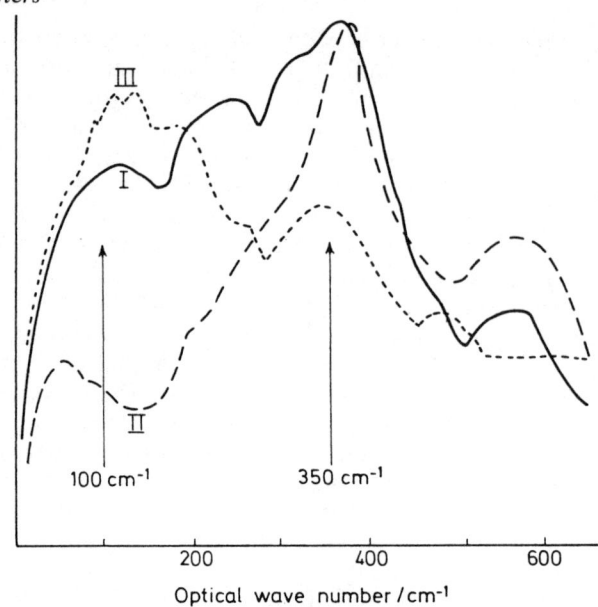

Figure 11.8 Comparison of the extrapolated neutron intensity curves for swollen versions of I, poly(methyl methacrylate); II, poly(methyl methacrylate-COOCD$_3$); and III, poly(α-chloromethyl methacrylate) (From Higgins, Allen and Brier[10], by courtesy of Polymer)

the results were extrapolated to a zero value of 2θ to avoid small effects due to dispersion and to give improved accuracy from the averaging process. The ordinate which is plotted in *Figures 11.7* and *11.8* is a generalized frequency function obtained from this extrapolation and is directly comparable with optical spectra.

In the *rubber* state of polymeric materials the polymer chains undergo rapid and continuous random changes in conformation, though there is no diffusion of the centre of mass of the polymer molecule. This has been studied by Allen *et al.*[11] from observations of the quasi-elastic peak in the neutron scattering pattern of polydimethylsiloxane. This peak is familiar in neutron studies of liquids where the elastic peak, normally associated with a solid, is, in an energy-analysis spectrum, broadened because of the transfer of energy from the neutron to the free atoms in a liquid. This is equivalent to the well-known Doppler broadening in optics. It may also occur in rubber-like materials where random motion of a diffusional nature may occur and the broadening may be analysed in terms of a diffusion coefficient D, using the relation

$$\Delta\omega = 2\hbar DQ^2 \quad (11.1)$$

which was developed by Larsson[12] for liquids. *Figure 11.9* shows the broadening for both the four-membered chain and ring forms of polydimethylsiloxane which are respectively

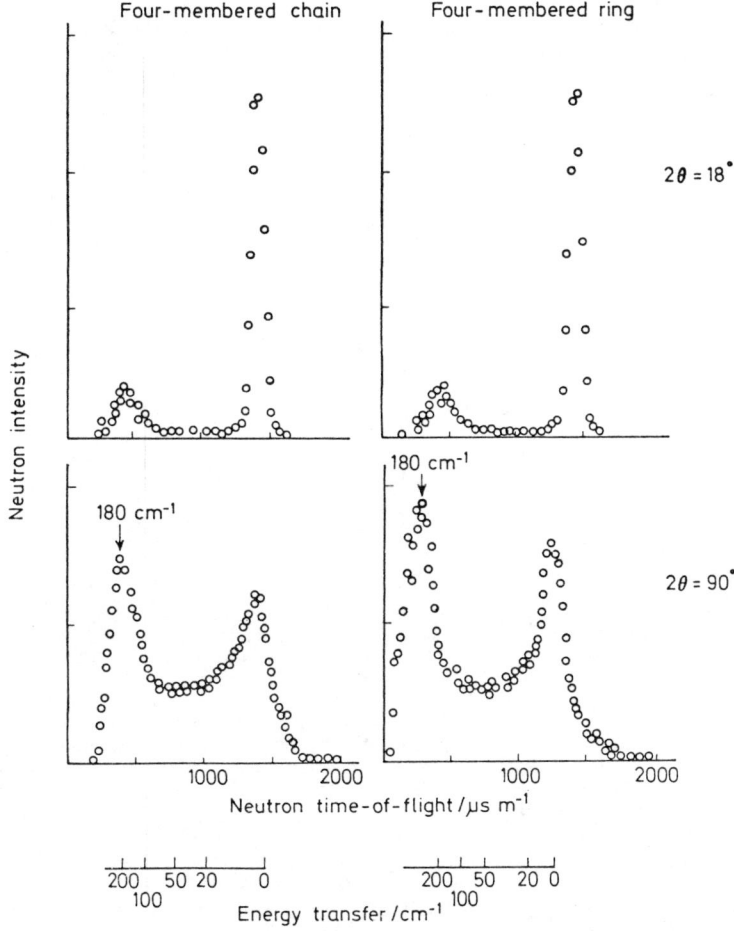

$$\text{CH}_3-(\text{Si}-\text{O})_3-\overset{\overset{\text{CH}_3}{|}}{\underset{\underset{\text{CH}_3}{|}}{\text{Si}}}-\text{CH}_3 \quad \text{and} \quad \overset{\overset{\text{CH}_3}{|}}{\underset{\underset{\text{CH}_3}{|}}{-(\text{Si}-\text{O})_4-}}$$

These particular materials were chosen for study because they have extremely flexible molecules and comparatively low values for the bulk viscosity. It is evident from the figure how the broadening of the quasi-elastic peak increases with the value of Q, i.e. $4\pi \sin \theta/\lambda$, in accordance

Figure 11.9 Time-of-flight spectra for, respectively, the four-membered chain and ring forms of polydimethylsiloxane, with incident neutrons of wavelength 4.5 Å. The neutrons are observed at scattering angles 2θ of 18 and 90° in the upper and lower pairs of diagrams, respectively (From Allen[14], by courtesy of Oxford University Press)

148 *Polymers*

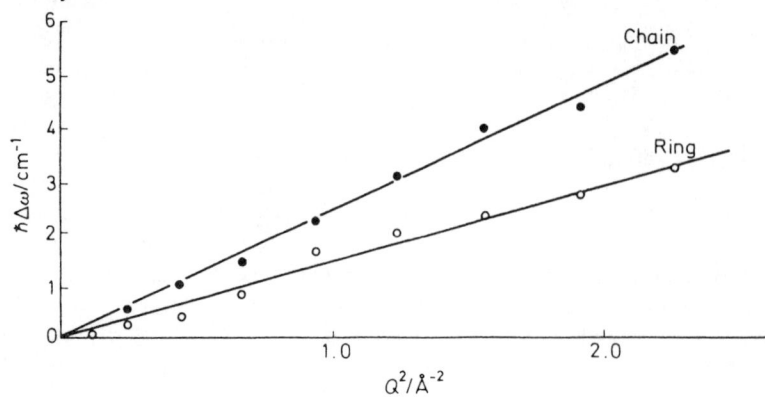

Figure 11.10 Determination of the self-diffusion coefficient D for ring and chain forms of polydimethylsiloxane, from measurements of quasi-elastic broadening using equation (11.1) (From Allen[14], by courtesy of Oxford University Press)

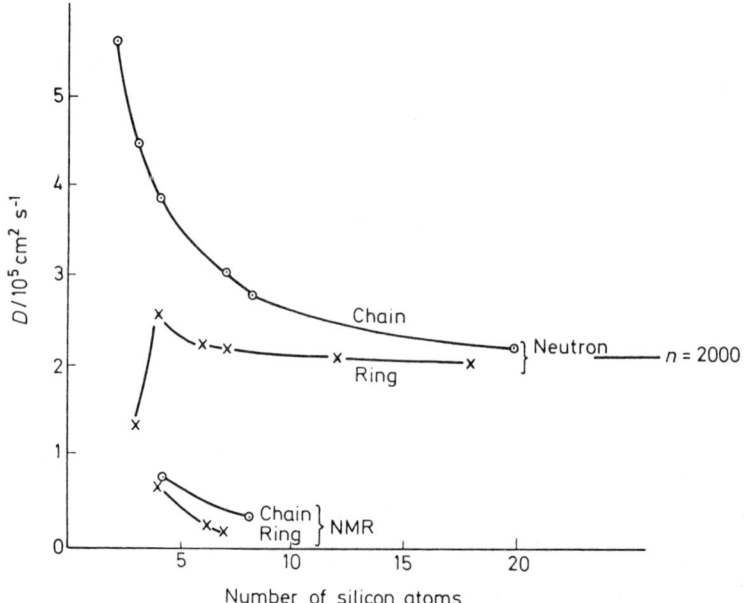

Figure 11.11 The variation of the self-diffusion coefficient D with the number of silicon atoms in ring and chain forms of polydimethylsiloxane, giving a comparison of the results from neutron scattering and NMR (From Allen et al.[11], by courtesy of the Chemical Society, London)

with the above equation. *Figure 11.10* indicates how this and similar data yield a value of D and shows that this is substantially smaller for the ring than for the chain compounds. However, examination of higher members of the two series, in which the number of silicon atoms is steadily increased from 4 to 2000, indicates that little difference persists between chain and ring (*Figure 11.11*) and that the values of D are greater, by two orders of magnitude, than the translational diffusion coefficients determined from n.m.r. spin-echo measurements. The independence of length of ring or chain suggests that the motion which is observed by the neutrons is internal rotation of short sequences of repeat units in a diffusive manner. Further experiments by Allen, Higgins and Wright[13] on other materials, such as poly(propylene oxide), in selectively deuterated forms, and poly(ethylene oxide), have confirmed this suggestion, disposing in particular of any possibility that the motion was associated with side groups of the main chain.

References

1. TASUMI, M. and SHIMANOUCHI, T., *J. Chem. Phys.,* **43,** 1245 (1965)
2. BOUTIN, H., PRASK, H., TREVINO, S.F. and DANNER, H., in *Inelastic Scattering of Neutrons,* Vol.II, 407, I.A.E.A., Vienna (1965)
3. MYERS, W.J., SUMMERFIELD, G.C. and KING, J.S., *J. Chem. Phys.,* **44,** 184 (1965)
4. KITAGAWA, T. and MIYAZAWA, T., *J. Chem. Phys.,* **47,** 337 (1967)
5. TWISTLETON, J.F. and WHITE, J.W., in *Neutron Inelastic Scattering,* 301, I.A.E.A., Vienna (1972)
6. FELDKAMP, L.A., VENKATARAMAN, G. and KING, J.S., *Neutron Inelastic Scattering,* Vol.II, 159, I.A.E.A., Vienna (1968)
7. TWISTLETON, J.F. and WHITE, J.W., *Polymer,* **13,** 40 (1972)
8. LAGARDE, V., PRASK, H. and TREVINO, S., *Discuss. Faraday Soc.,* **48,** 15 (1969)
9. HOLLIDAY, W. and WHITE, J.W., *Pure Appl. Chem.,* **26,** 545 (1971)
10. HIGGINS, J.S., ALLEN, G. and BRIER, P.N., *Polymer,* **13,** 157 (1972)
11. ALLEN, G., BRIER, P.N., GOODYEAR, G. and HIGGINS, J.S., *Faraday Symp. Chem. Soc.,* **6,** 169 (1972)
12. LARSSON, K.-E., in *Neutron Inelastic Scattering,* Vol.I, 397, I.A.E.A., Vienna (1968)
13. ALLEN, G., HIGGINS, J.S. and WRIGHT, C.J., *J. Chem. Soc. Faraday Trans. II,* **70,** 348 (1974)
14. ALLEN, G., in *Chemical Applications of Thermal Neutron Scattering,* ed. B.T.M. Willis, 97, Oxford University Press (1973)

12

LIQUIDS, GLASSES AND GASES

We shall see that neutrons have a number of distinctive advantages over X-rays for the study of liquids and glasses. Most of these only become evident from a detailed examination of the scattering processes in each case but, for liquids in particular, there are a number of very practical considerations which may be appreciated more readily. Thus the low absorption coefficients for neutrons means that much larger samples may be examined, giving results which are free from uncertainties concerning the contamination of a liquid surface by foreign atoms. It is also much easier to use neutrons for measurement at high temperatures, studying liquid metals and alloys for example, and under pressure. This too arises because of the low absorption coefficients which permit the use of strongly built containers of adequate strength.

12.1 Liquids

From the point of view of the atomic structure the main characteristic of a liquid is that the long-range atomic order, which is so characteristic of a crystalline solid, does not exist. An immediate consequence of this is that the well-defined Bragg peaks which are so prominent in the diffraction pattern of a solid, and which are extremely narrow in angular width because they arise from coherent superposition of scattered contributions over a distance of the order of 1000 Å, are not present in the pattern of a liquid. As illustrations we show in *Figure 12.1* neutron diffraction patterns for liquid lead and water. The pattern of the former, which is of course monatomic, can be interpreted straightforwardly, though not particularly easily. Water is not a simple liquid and its structure depends on molecular orientation and association and for this and other reasons, to which we shall refer later, the interpretation of its neutron diffraction pattern is considerably more complicated and, in fact, has not yet been properly achieved. However, the main feature of the pattern for water in the figure is that such detail as might otherwise be expected is completely swamped by the incoherent scattering from the hydrogen atoms. The absence of the Bragg peaks in the liquid patterns means that it is no longer possible, at least in an approximate way, to separate elastic from inelastic scattering and indeed there is no strictly elastic scattering from a liquid, since the atoms are not attached to specific sites but are free to move. As a result of this diffusion motion there

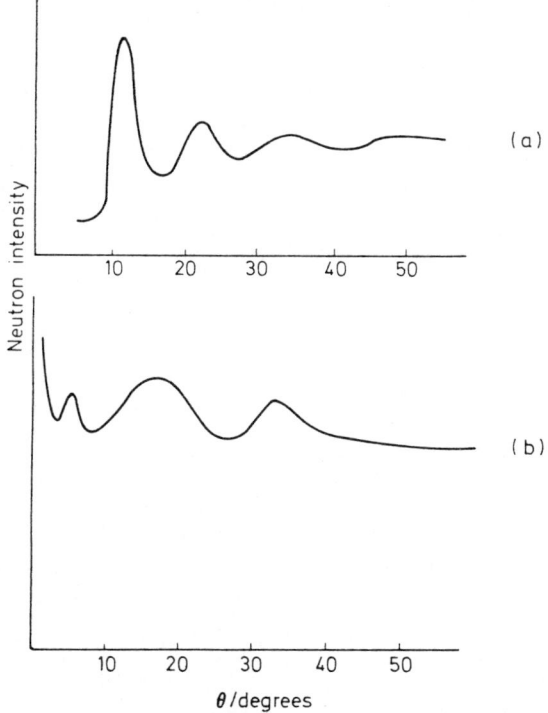

Figure 12.1 Neutron diffraction pattern of liquid lead and liquid water for neutrons of wavelength about 1.1 Å

are inevitably small exchanges of energy with the incoming neutrons and the scattering is, at the best, only 'quasi-elastic'. At the same time much more substantial exchanges of energy may take place, with the excitation or de-excitation of molecular rotations and vibrations of atoms within molecules. This is emphasized in *Figure 12.2*, which shows a typical energy analysis, in the form of a wavelength-distribution curve, for the neutrons scattered by a liquid. The pattern divides essentially into two parts. At the right-hand side the wavelength spread of the incident neutron beam is broadened by the quasi-elastic scattering, whereas the peaks on the left of the diagram arise from genuine inelastic-scattering exchanges.

The structural features and considerations which we have just outlined lead to the conclusion that the interpretation of the scattering by liquids can only be made successfully in terms of the general space–time correlation function of Van Hove[1]. $G(r,t)$ is a function of space and time which gives the probability of finding an atom at distance r from an origin at a time t later than some zero of time, assuming that an atom had been observed at the origin when $t = 0$. This extremely complicated function, which is far from being known in detail at present, can be divided into two parts. First, the atom which we observe at r and time t may be the *same* one which we saw at $r = 0$ when $t = 0$, and

152   Liquids, Glasses and Gases

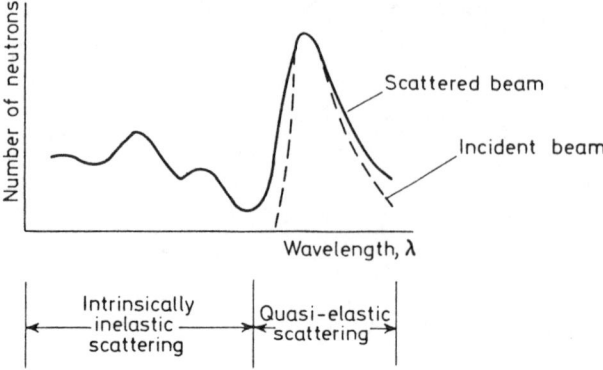

Figure 12.2   A wavelength analysis of the neutrons observed at a fixed angle of scattering, for a typical liquid. The pattern divides into two parts, the quasi-elastic scattering which depends on atomic diffusion and the true inelastic scattering which is determined by the structure

we call this part of the probability function the 'self-correlation function' $G_s(r,t)$; on the other hand, we may observe a different atom at r,t and this part of the function is termed the 'distinct-correlation function' $G_d(r,t)$. In principle the neutron scattering pattern contains information from which $G_s(r,t)$ and $G_d(r,t)$ can be separately deduced. That this should be so follows from the fact that the wave field of the scattered neutrons will depend on the space-time history of the sample. Thus a neutron of velocity v detected at a point distant r from the sample at time t will have been scattered at a time near $t - r/v$. Moreover, if inelastic scattering has taken place then energy will have been transferred to atomic nuclei and there will be systematic correlation between atomic positions at different times.

If we perform a scattering experiment with monochromatic neutrons of energy E_0 then the most refined measurement would measure $d^2\sigma/d\Omega dE$, the scattered intensity at each angle 2θ in terms of the number of neutrons per unit solid angle Ω and per unit energy range, remembering that the scattered neutron will have energy E, where $E_0 - E = \hbar\omega$ as defined in equation 3.2. It was shown by Van Hove that for the scattering by a group of N nuclei

$$\frac{d^2\sigma}{d\Omega dE} = N \frac{\kappa}{\kappa_0} \frac{1}{2\pi\hbar} \int_{-\infty}^{\infty} dt \exp(-i\omega t) \sum_r \exp(iQ\cdot r)[(\bar{b})^2 G(r,t) +$$

$$\{(\overline{b^2}) - (\bar{b})^2\} G_s(r,t)] \quad (12.1)$$

and this expression separates into two parts which are distinguished as 'coherent' and 'incoherent' differential scattering cross-sections. Thus

$$\frac{d^2\sigma_{coh}}{d\Omega dE} = N \frac{\kappa}{\kappa_0} \frac{1}{2\pi\hbar} (\bar{b})^2 \int_{-\infty}^{\infty} dt \exp(-i\omega t) \sum_r \exp(iQ\cdot r) G(r,t) \quad (12.2)$$

and

$$\frac{d^2\sigma_{incoh}}{d\Omega dE} = N\frac{\kappa}{\kappa_0}\frac{1}{2\pi\hbar}\{(\overline{b^2}) - (\overline{b})^2\}\int_{-\infty}^{\infty} dt \exp(-i\omega t)\sum_r \exp(i\mathbf{Q}\cdot\mathbf{r})G_s(r,t) \quad (12.3)$$

where the incoherence is brought about by the effects of nuclear spin and different isotopes, which produce a distinction between the values of $\overline{b^2}$ and $(\overline{b})^2$, as we discussed in Chapter 2.

The two integrals which appear in these expressions are normally described as 'scattering laws' and are written as $S(Q,\omega)$ and $S_s(Q,\omega)$, being functions of both the momentum transfer Q and the energy interchange $\hbar\omega$. Thus

$$\text{coherent } S(Q,\omega) = \frac{1}{2\pi}\iint G(r,t)\exp\{i(\mathbf{Q}\cdot\mathbf{r} - \omega t)\}dr\,dt \quad (12.4)$$

and

$$\text{incoherent } S_s(Q,\omega) = \frac{1}{2\pi}\iint G_s(r,t)\exp\{i(\mathbf{Q}\cdot\mathbf{r} - \omega t)\}dr\,dt \quad (12.5)$$

and, further,

$$\frac{d^2\sigma_{coh}}{d\Omega dE} = N\frac{(\overline{b})^2}{\hbar}\frac{\kappa}{\kappa_0}S(Q,\omega) \quad (12.6)$$

and

$$\frac{d^2\sigma_{incoh}}{d\Omega dE} = \frac{N}{\hbar}\frac{\kappa}{\kappa_0}\{(\overline{b^2}) - (\overline{b})^2\}S_s(Q,\omega) \quad (12.7)$$

The term 'scattering law' is not completely satisfactory, for these expressions depend solely on the dynamics of the scatterer and not on the properties of the incident radiation.

In measuring the neutron scattering pattern of a liquid we record at each angle of scattering 2θ the total number of neutrons which enter the counter. It is more difficult to evaluate this number than it may appear at first sight: the difficulty may be understood best by considering the corresponding case in X-ray scattering. For X-rays the changes $\hbar\omega$ in energy brought about by inelastic scattering are negligible in comparison with the incident energy E_0 and, as *Figure 12.3* shows, all X-rays entering a detector at a given value of 2θ will have effectively the same value of both Q and ω. Accordingly for X-rays the total count will be

$$\int \frac{d\sigma}{d\Omega dE}dE = N(\overline{b})^2\int S(Q,\omega)\,d\omega \quad (12.8)$$

$$= N(\overline{b})^2 S(Q)$$

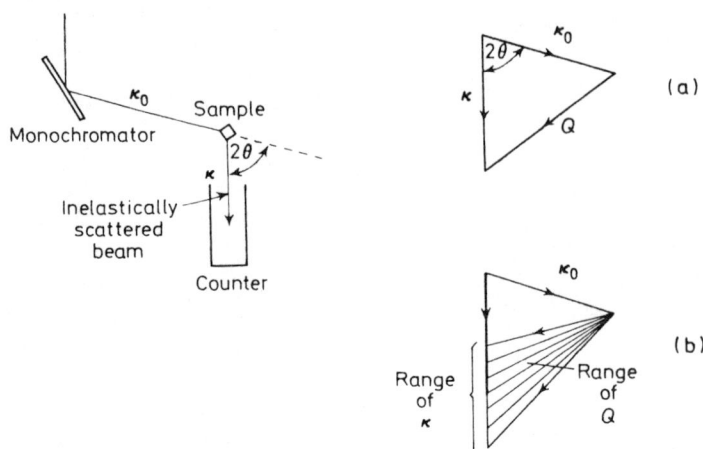

Figure 12.3 Diagrams of inelastic scattering. In the momentum diagram of (a) for X-rays, κ is effectively constant and equal to κ_0, so that Q is uniquely defined. For neutrons, (b), κ may be substantially less or greater than κ_0, giving a wide range of values for Q

where we term $S(Q)$ the 'structure factor' of the liquid, which can be shown by integration of equation 12.4 to be given by

$$S(Q) = 1 + \frac{4\pi\rho}{Q} \int_0^\infty \{g(r) - 1\} r \sin Qr \, dr \qquad (12.9)$$

where $g(r)$ is the value of $G(r,t)$ when $t = 0$ and is therefore a snapshot picture of the atomic order in the liquid and is often described as the 'static approximation'; ρ is the number of atoms per unit volume. Then by Fourier inversion it follows that

$$g(r) = 1 + \frac{1}{2\pi^2 \rho r} \int_0^\infty Q\{S(Q) - 1\} \sin Qr \, dQ \qquad (12.10)$$

On the other hand, for neutrons it is not possible to determine $S(Q)$ directly from the experiment since, as *Figure 12.3* shows, the neutrons entering the counter at a given angle 2θ vary over a considerable range of Q because the change $\hbar\omega$ may be a considerable fraction of E_0. Moreover, the efficiency of the detector varies significantly with neutron energy. The result is that a number of corrections (the so-called Placzek[2] corrections) have to be made before the experimental neutron scattering patterns can be used to produce $S(Q)$ and, subsequently, $g(r)$. It is also to be noted that it is the *coherent* neutron scattering which can be used, via equation 12.10, to give $g(r)$ in this way. Accordingly the most accurate experiments with neutrons were done by Enderby and his colleagues with liquid metals such as lead (North et al.[3,4]) which are not only simple monatomic systems but also give a negligible amount of incoherent scattering. *Figure 12.4* shows a curve for $g(r)$ for liquid lead.

Liquids, Glasses and Gases 155

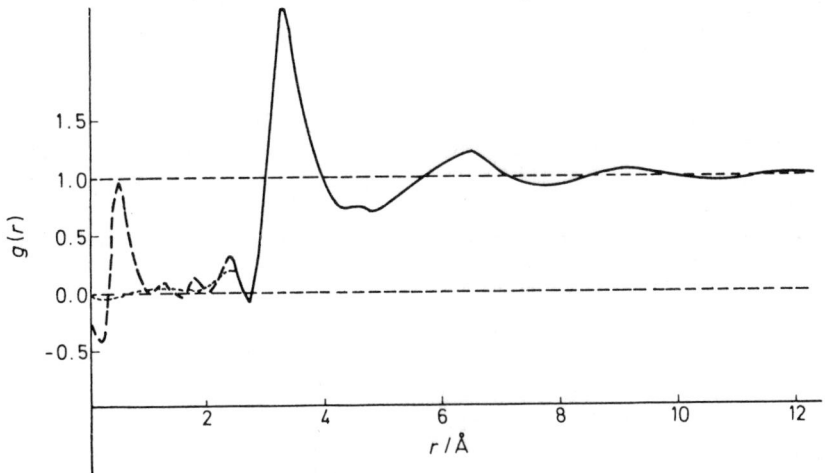

Figure 12.4 *An experimentally determined curve of g(r) for liquid lead at 873 K. The rapid rise below r = 1 and the subsequent ripples are spurious and when corrections are made the dotted curve is obtained for the region of r < 2 Å (After North et al.[4], by courtesy of J. Phys. C)*

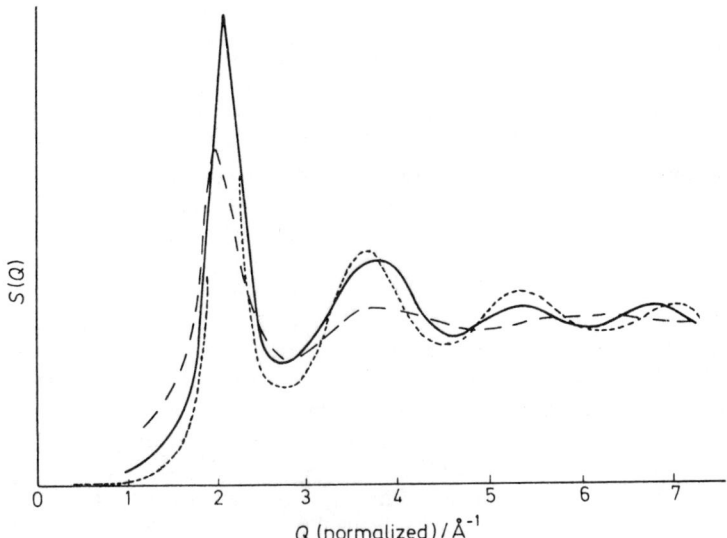

Figure 12.5 *A comparison of the structure factors S(Q) measured experimentally for liquid argon (——) and liquid rubidium (- - -), with calculations (-----) for a hard-sphere model (After Page et al.[5], by courtesy of Phys. Lett.)*

The accuracy of the Fourier inversion is restricted by the cut-off which occurs experimentally at both large and small values of Q. An alternative way of assessing the experimental data is to compare the results with calculations made for various models and *Figure 12.5* shows such a comparison, using the hard-sphere model of a liquid, for liquid argon and rubidium (Page et al.[5]).

12.1.1 BINARY ALLOYS

The real virtue of using neutrons for studies of liquids emerges when we extend our consideration beyond the simple monatomic liquid metals to, first, binary alloys and then, more speculatively, to more complicated liquids. In the case of a binary liquid AB we have to consider three different interatomic potentials, related to the interactions A⋯A, A⋯B and B⋯B, and for each of them we can define a partial structure-factor of the form

$$S_{AB}(Q) = 1 + \frac{4\pi\rho}{Q} \int_0^\infty \{g_{AB}(r) - 1\} r \sin Qr \, dr \qquad (12.11)$$

where ρ remains equal to the *total* number of atoms per unit volume and $g_{AB}(r)$ is the probability that there is an atom of type B at position r when there is an atom A at the origin. The equation for the differential scattering cross-section then becomes

$$\frac{d\sigma_{coh}}{d\Omega} = N(c_A b_A^2 + c_B b_B^2) + N\{c_A^2 b_A^2 (S_{AA} - 1) + c_B^2 b_B^2 (S_{BB} - 1) + 2c_A c_B b_A b_B (S_{AB} - 1)\} \qquad (12.12)$$

where c_A and c_B are the atomic fractions of the two component atoms A and B and b_A and b_B are their coherent scattering lengths; N is the total number of atoms in the sample. This expression makes clear that the neutron intensity at any particular angle of scattering, or for a given value of Q, will depend on the corresponding values of *three* partial structure factors S_{AA}, S_{BB} and S_{AB}. Using X-rays it will not be possible to determine these separately but neutrons can exploit the advantages of manufacturing and measuring alloys from different isotopes with — in favourable cases — different values of scattering lengths. If three different isotopes are available then there will be three different equations from which to determine the three variables. Thus Enderby, North and Egelstaff[6] prepared the alloy Cu_6Sn_3 in turn from natural copper, ^{63}Cu and ^{65}Cu, for which the scattering lengths are 0.76, 0.67 and 1.11 × 10^{-12} cm, respectively. Similarly Page and Mika[7] studied four samples of molten CuCl consisting of $^{63}Cu^{35}Cl$, $^{nat}Cu^{nat}Cl$, $^{65}Cu^{nat}Cl$ and $^{65}Cu^{37}Cl$, for which the ratio of the scattering lengths of copper and chlorine are 0.58, 0.82, 1.16 and 2.41, respectively. It must be emphasized that the extraction from the intensity data of the separate curves for the partial structure factors requires experimental measurements of high accuracy and the difference in amplitude between the isotopes must be large enough to make the solution of the equations as insensitive as possible to experimental errors.

The $g(r)$ curves derived by Fourier inversion of the $S(Q)$ data for CuCl are reproduced in *Figure 12.6* and include the usual spurious oscillations at low values of r which arise through the terminations of the range of

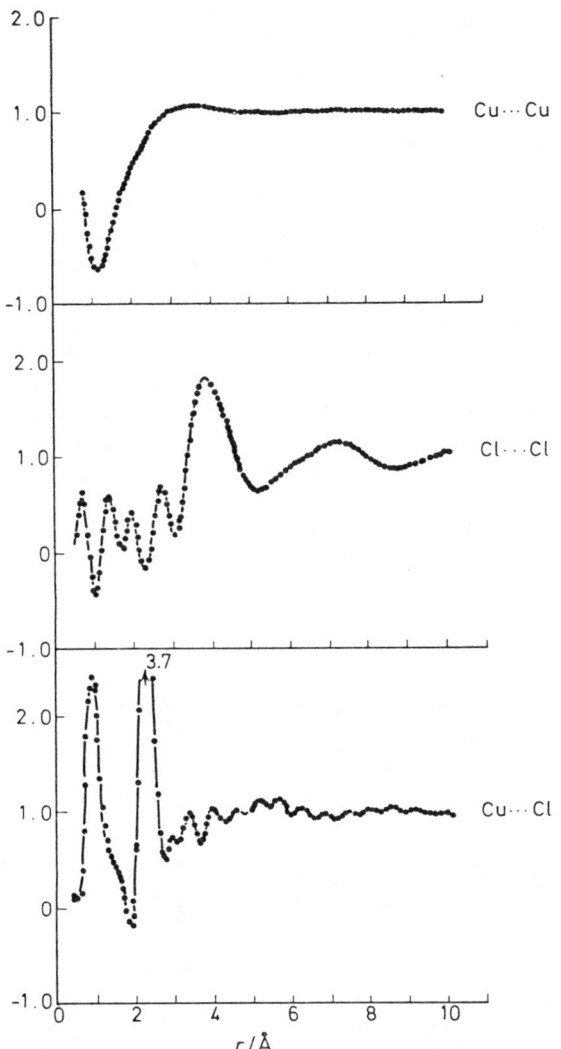

Figure 12.6 The three partial radial-distribution functions g(r) for liquid CuCl (From Page and Mika[7], by courtesy of J. Phys. C)

Q. For Cl⋯Cl, $g(r)$ is similar to what is observed in monatomic liquids, but Cu⋯Cu is almost featureless and Cu⋯Cl has a single very sharp peak at 2.3 Å, if we ignore the peak at 0.9 Å which is almost certainly artificial. Page and Mika discuss these curves in relation to the state of ionization of the salt and favour a model in which the CuCl is not fully ionized, having a large number of groups in which several Cl$^-$ ions associate with one Cu$^+$ ion and necessarily an excess of itinerant copper ions. However, Powles[8] has shown that it is equally possible to interpret the results solely in terms of CuCl molecules having an internuclear separation of 2.3 Å and some correlation between the orientations of neighbouring molecules.

12.1.2 AQUEOUS SOLUTIONS

Studies of aqueous solutions have been made by Narten, Vaslow and Levy[9] and by Enderby, Howells and Howe[10]. The former studied lithium chloride solutions in concentrations of up to one molecule for three molecules of water, giving a saturated solution. Both X-rays and neutrons were used, employing heavy water and the lithium-7 isotope for the neutron measurements. Comparisons of the patterns obtained with the two radiations are informative because of the different ratios of the scattering amplitudes of Li, Cl, H(D) and O. Figures 12.7(a) and (b) show the pair correlation functions derived from the diffraction patterns for X-rays and neutrons. For X-rays the pattern for weak solutions is dominated by the near-neighbour O–O interaction with a peak at 2.85 Å; for more concentrated solutions the large scattering amplitude of chlorine

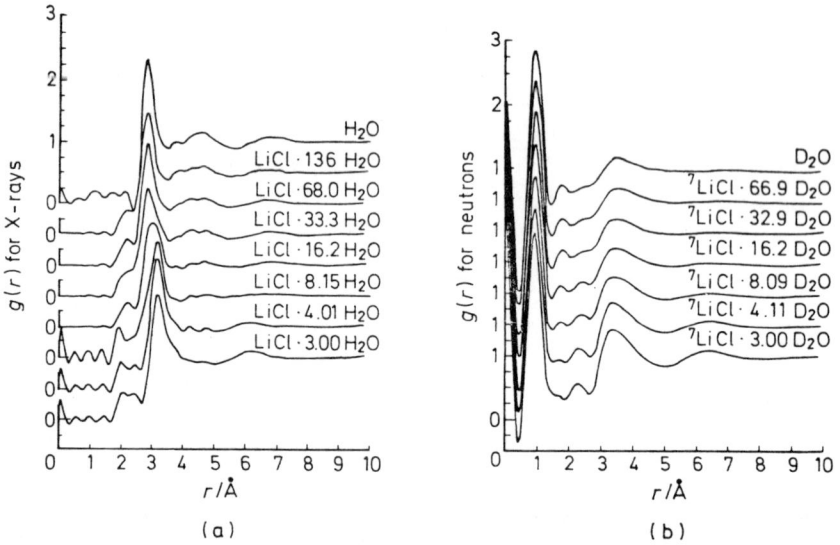

Figure 12.7 The pair-distribution functions g(r) measured with (a) X-rays and (b) neutrons for aqueous solutions of lithium chloride of various concentrations. The individual curves are displaced vertically to avoid overlap (From Narten, Vaslow and Levy[9], by courtesy of J. Chem. Phys.)

leads to a significant O–Cl peak at 3.2 Å. For neutrons, interactions which involve the more heavily scattering deuterium atom dominate the pattern. Thus the main peak at 0.94 Å is accounted for by the O–D intramolecular separation. The following peak at about 1.7 Å is a combination of an *intra*molecular D–D separation at 1.5 Å and an *inter*molecular O–D contact, via a hydrogen bond, at 1.9 Å. For the more

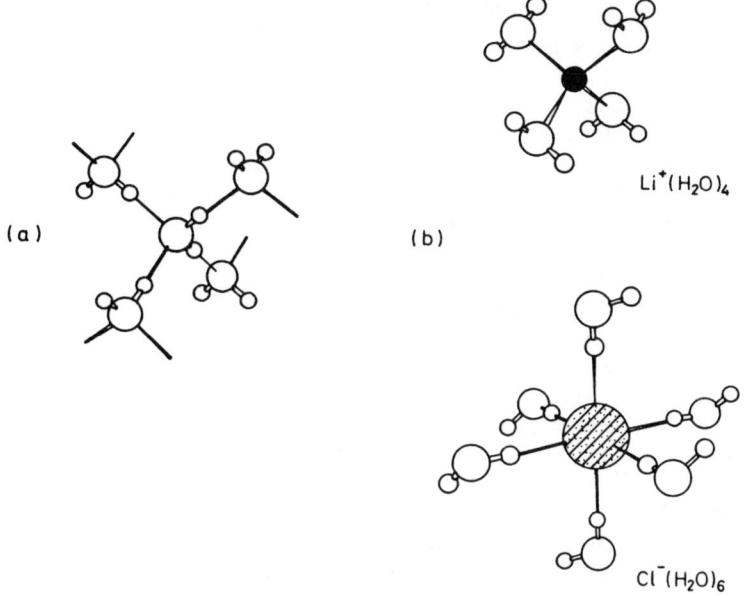

Figure 12.8 Diagram (a) shows the environment of water molecules in pure water and dilute LiCl solutions. The diagrams (b) show the four-fold and six-fold coordination by water molecules of the Li^+ and Cl^- ions respectively in concentrated solutions (From Narten, Vaslow and Levy[9], by courtesy of J. Chem. Phys.)

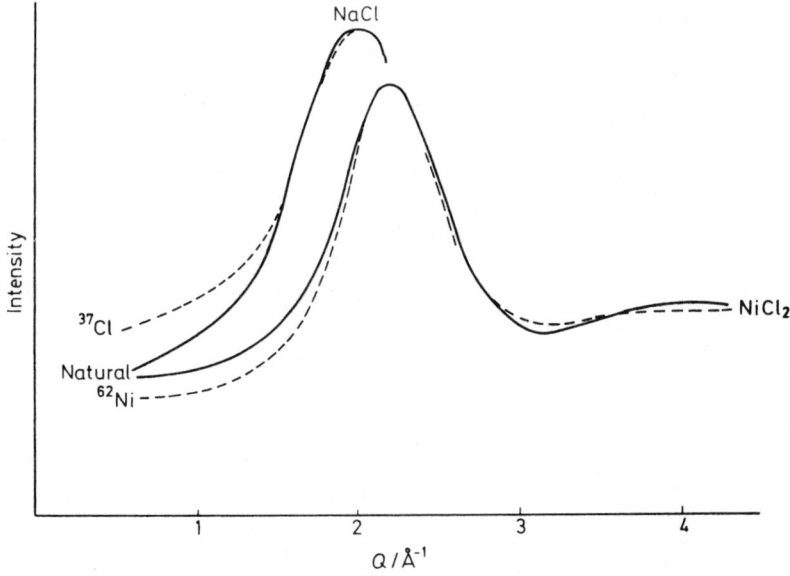

Figure 12.9 Evidence that the neutron scattering pattern of aqueous solutions depends on the isotopic content of the solute, for NaCl (prepared from natural Cl and ^{37}Cl) and $NiCl_2$ (prepared from natural Ni and ^{62}Ni) (From Enderby, Howells and Howe[10], by courtesy of Chem. Phys. Lett.)

concentrated solutions the latter component of the peak disappears as it becomes less and less likely that water molecules will be adjacent to each other. The patterns for the most concentrated solutions were analyzed by least-squares in terms of hydrated ions of the form $Cl^-(H_2O)_m$ and $Li^+(OH_2)_n$, leading to satisfactory fits with four-fold coordination of the lithium ion and six-fold coordination of the chlorine ion. These coordinations, both in pure water and dilute solutions (*a*) and in the concentrated solutions (*b*), are illustrated in *Figure 12.8*.

Enderby, Howells and Howe[10] have applied the technique of isotopic enrichment to determine values for the partial structure factors relating to A···A, A···B and B···B for solutions of $NiCl_2$, NaCl and $BaCl_2$ in heavy water. This technique promises to be as valuable for the study of solutions as for the alloys and molten salts which we examined earlier. *Figure 12.9* shows the distinctive differences for the different solutions and isotopes, and from curves of this kind the individual partial structure-factor curves can be obtained in the following way. For an electrolyte consisting of a salt MX_n dissolved in heavy water it can be shown (Howe, Howells and Enderby[11]) that the neutron scattering $F(Q)$ depends on 10 partial structure factors, including S_{MM} the factor for the metal ions. If separate measurements are made in which the metal is present in three different isotopic forms, with substantially different scattering amplitudes for neutrons, then, in principle, the value of S_{MM} can be deduced. For $NiCl_2$ the above authors used, in turn, material manufactured from natural nickel, ^{62}Ni and a mixture of ^{60}Ni and ^{62}Ni which has a zero value of coherent scattering amplitude. It can be shown that

$$S_{NiNi} = 1 + \frac{b_{62}(F_N - F_0) - b_N(F_{62} - F_0)}{c_2^2 b_{62} b_N (b_N - b_{62})} \quad (12.13)$$

where F_N, F_{62} and F_0 are the values of $F(Q)$ for the samples of 'natural', 62 and 'zero scattering' nickel, respectively; b_N and b_{62} are the scattering amplitudes of natural nickel and ^{62}Ni and c_2 is the atomic concentration of the nickel cations in the solution. The measurements, including the application of the various necessary corrections, could be carried out sufficiently accurately to extract the data shown in *Figure 12.10*, which gives the variation with Q of S_{NiNi} for solutions which contain, respectively, 4.7, 3.5, 3.1 and 2.5 mole % concentrations, given by Howe, Howells and Enderby[11]. In comparison with curves for alloys and molten salts these curves are remarkable for the values of the order of 60 which the structure factor attains for the prominent peak: a typical value for an alloy is only 2. At the same time the oscillating features of the patterns remain clear at high values of Q. This noteworthy result suggests that there is a high degree of long-range order among the nickel ions and that the nickel ions in these concentrated solutions are arranged on what approaches a regular lattice. The peak near a Q value of 1 Å^{-1} in S_{NiNi} indicates a peak at about 6 Å in the interatomic separation, since $d = \lambda/2 \sin \theta = 2\pi/Q$, and the existence of such a peak in the

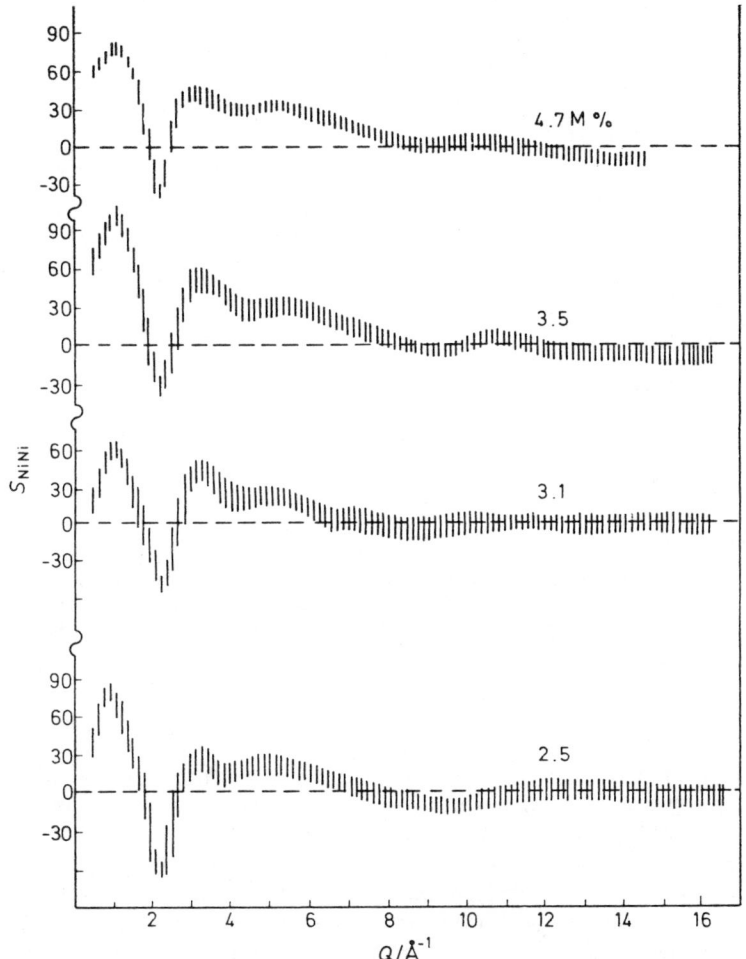

Figure 12.10 The variation with Q of the partial structure factor $S_{Ni\text{-}Ni}$ for aqueous solutions of $NiCl_2$ of various percentage compositions (From the data of Enderby and Howe)

neutron scattering pattern has been detected by Neilson, Howe and Enderby[12] in solutions varying in molar concentration from 2.78 to 6.61%. Their results are illustrated in *Figure 12.11(a)* for the two extremes of concentration. Below 2.78% the peak becomes weak; the maximum value of 6.61% corresponds to saturation. Moreover, the angular position of the peak varies with concentration in such a way that the reciprocal of the corresponding interatomic separation is proportional to the cube-root of the concentration, thus providing strong evidence in favour of a quasi-lattice of nickel ions. These results are shown at (b) in *Figure 12.11* and the observations are being extended to lower concentrations in order to explore the manner in which this peak eventually disappears.

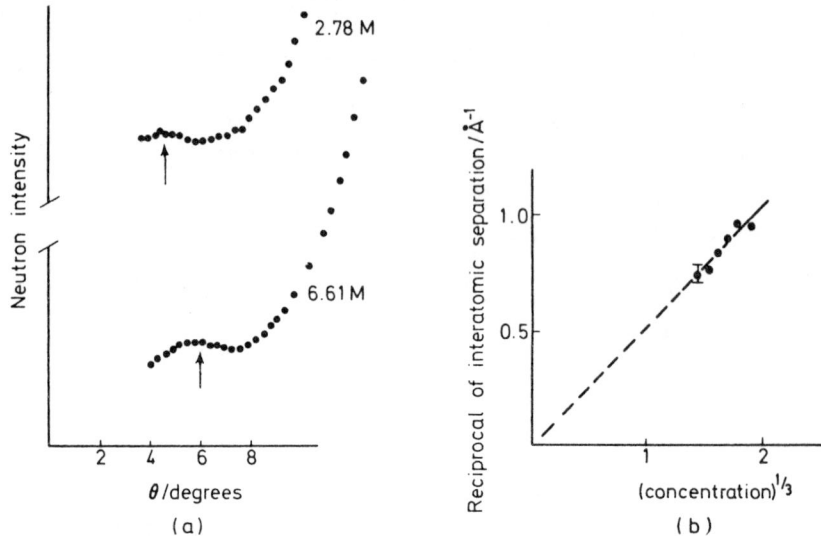

Figure 12.11 The 'quasi-lattice' peak of aqueous solutions of $NiCl_2$. Diagram (a) shows the measured diffraction patterns for solutions of strengths 2.78 and 6.61 M and (b) shows that the reciprocal of the interatomic separation, obtained from the angular position of the peak, is proportional to the cube-root of the concentration (From Neilson, Howe and Enderby[12], by courtesy of Chem. Phys. Lett.)

12.1.3 LIQUID WATER

In comparison with the liquid metals and liquid binary alloys which we have just discussed, the determination of the structure of liquid water is a very different problem. For ordinary light water the determination of the *structure*, as distinct from the diffusive atomic motion which we shall consider later in this chapter, is practically impossible at the present time because of the very large incoherent scattering from ordinary hydrogen. With *heavy* water the scattering is mainly coherent but it is still difficult to deduce $S(Q)$, and subsequently $g(r)$, from the scattering patterns with much accuracy owing to uncertainty in estimating the Placzek correction, which is difficult to apply in the case of light elements. Even when $S(Q)$ has been deduced there remains a substantial problem in interpreting it since it is necessary to take into account not only the structure of the water molecules themselves but also the relative orientation which exists between neighbouring molecules. The forces between molecules will not be spherically symmetrical, as they are between atoms in simple liquid metals and binary alloys. Page and Powles[13] have discussed both the X-ray and neutron scattering data for heavy water, comparing the experimental information with calculations for both unoriented molecules and various postulated models of orientation, with the aid of the concept of the 'molecular-centres structure-factor' $S_c(Q)$, described by Egelstaff *et al.*[14] $S_c(Q)$ describes, in terms of equation 12.9, the distribution within the liquid of the centres of the molecules and can be shown to be related to the observed structure-factor $S_m(Q)$ by the equation

$$S_m(Q) = f_1(Q) + f_2(Q)[S_c(Q) - 1] \tag{12.14}$$

where $f_1(Q)$ is a form-factor appropriate to a single isolated molecule and $f_2(Q)$ depends on the particular correlation of orientation assumed between neighbouring molecules. Both $f_1(Q)$ and $f_2(Q)$ can be calculated for a particular model. Moreover, if it is assumed that the scattering centres of the molecules have the same distribution as the oxygen atoms, it will be expected that the observed X-ray scattering will give a direct measure of $S_c(Q)$, since the deuterium atoms will give little X-ray scattering. Accordingly from equation 12.14 a value of $S_m(Q)$ can be derived for comparison with the neutron data. The results of this procedure carried out by Page and Powles are summarized in *Figure 12.12*, which compares

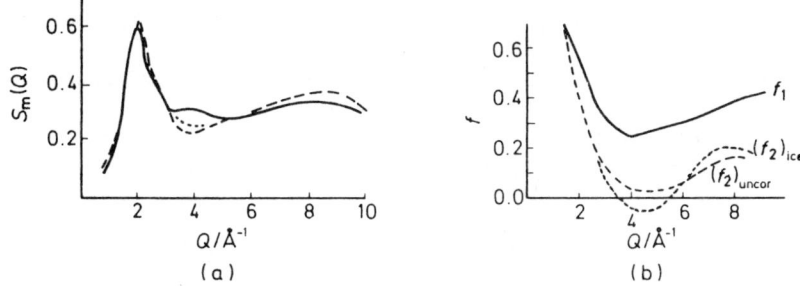

Figure 12.12 X-ray and neutron scattering data for heavy water. In (a) the observed neutron scattering function $S_m(Q)$ (———) is compared with values expected if the X-ray molecular-centres scattering function is assumed to be correct and the water molecules are assumed, in turn, to be either uncorrelated (curve ---) or 'icy' (curve -----). The distinction between these latter two curves arises from the substantial difference between the values of the form-factor f_2 for the uncorrelated and 'icy' models: the appropriate form-factor curves are drawn at (b) (After Page and Powles[13], by courtesy of Mol. Phys.)

the neutron-determined curve for $S_m(Q)$ with derived curves for uncorrelated molecules and for a model with the orientation as in ice. The agreement is surprisingly good except in the region of $Q = 4$ Å^{-1}. In this latter region a correlated model is better than an uncorrelated one, but the authors were not able to postulate any correlation of orientation which gave agreement comparable with that observed over the rest of the range of Q.

The development, in recent years, of methods of polarization analysis promises much improvement in the studies of liquids containing hydrogen, by providing a means of separating the coherent and incoherent scattering. It can be shown from equation 2.10 that

$$\text{non-spin flip cross-section,} \quad \sigma_{++} = \frac{1}{3}\sigma_{\text{incoh}} + \sigma_{\text{coh}} \tag{12.15}$$

$$\text{spin-flip cross-section,} \quad \sigma_{+-} = \frac{2}{3}\sigma_{\text{incoh}} \tag{12.16}$$

164   Liquids, Glasses and Gases

so that the separate measurement of σ_{++} and σ_{+-} will give the individual values of σ_{coh} and σ_{incoh}. The success of this method depends on the availability of a high-flux reactor and future development of efficient polarizing crystals giving high reflectivities. Some preliminary measurements with heavy water have been made by Dore, Clarke and Wenzel[15].

12.1.4 ATOMIC MOTION

We have already seen that the coherent neutron scattering, which we can readily measure for liquid metals, depends on the total correlation function $G(r,t)$ and from a measurement of the scattering we deduce $g(r)$, which amounts to a snapshot picture of the liquid. On the other hand, the incoherent scattering, as equation 12.5 indicates, depends solely on the self-correlation function $G_s(r,t)$. This function, which describes motion in the liquid on an atomic scale, must be consistent with Fick's law of diffusion which describes the motion on a macroscopic scale. According to this law we know that when a concentration gradient dC/dx of molecules exists across a plane, then there will be a movement of matter J_x across the plane proportional to the concentration gradient. Thus

$$J_x = -D \frac{dC}{dx}$$

where D is the diffusion constant. It follows that for large distances and times the mean-square distance travelled in time t is given by

$$\overline{r_t^2} = 6\,Dt \qquad (12.17)$$

but this will not hold for short times, during which the accelerating forces have insufficient time to increase the velocity and for which r will effectively equal vt, where v is the atomic velocity and

$$\overline{r_t^2} = \overline{v^2}t^2 = 3k_B \frac{T}{M} t^2 \qquad (12.18)$$

where k_B is Boltzmann's constant and M is the atomic mass. The time at which equation 12.17 may be considered to commence to operate may plausibly be estimated as the time required by an atom to move a few atomic diameters. At room temperature a typical atomic velocity is 3×10^2 m s^{-1} and a distance of 3 Å will be covered in 10^{-12} s. Likewise a *thermal* neutron will spend a similar length of time in traversing such a distance and its scattering pattern will, in principle at least, produce a detailed picture of how $\overline{r_t^2}$ varies with time, since

$$\overline{r_t^2} = \int_0^\infty r^2 G_s(r,t) 4\pi r^2 \, dr \qquad (12.19)$$

Two methods have been used for interpreting the data. First, Brockhouse et al.[16] have measured at various angles of scattering the energy distribution of the scattered neutrons, as in *Figure 12.13*, thus yielding a curve

Liquids, Glasses and Gases 165

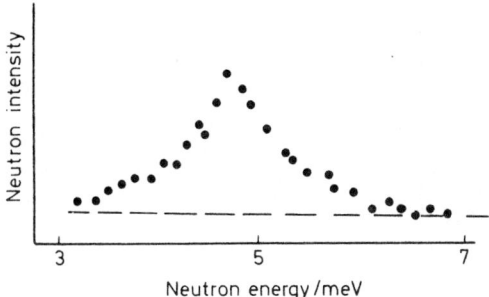

Figure 12.13 The quasi-elastic peak (see Figure 12.2) in the scattering pattern for liquid water at 300 K (From Brockhouse et al.[16], by courtesy of I.A.E.A.)

of (Q,ω) against ω. From a series of such curves it is possible to show how $G_s(r,t)$ varies with r,t by utilizing the relation

$$G_s(r,t) = \frac{1}{2\pi} \iint S_s(Q,\omega) \exp\{i(Q \cdot r - \omega t)\} \, dQ \, d\omega \qquad (12.20)$$

which is the Fourier transform of equation 12.5. Some results by Sakamoto et al.[16,17] are given in *Figure 12.14* and by utilizing equation 12.19 curves for the variation of r_t^2 with t can be constructed. These curves are shown in *Figure 12.15* from which it is seen that the motion is

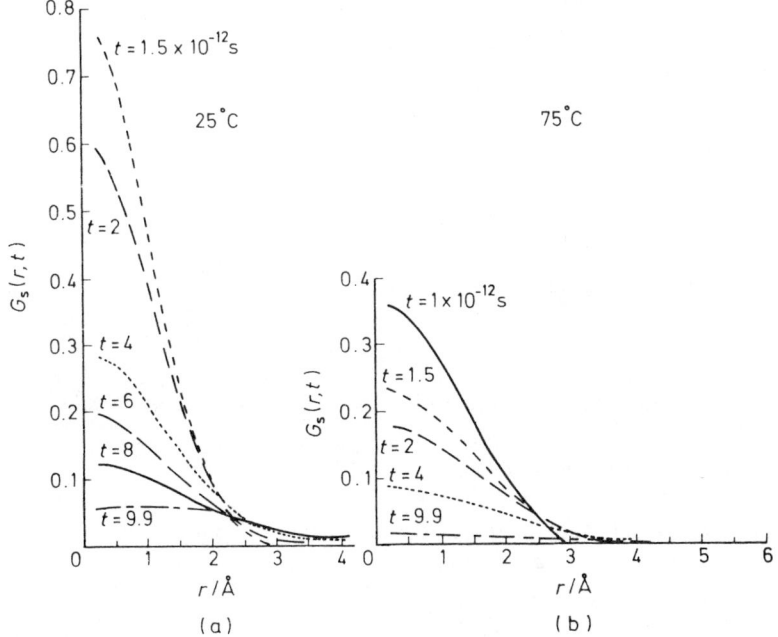

Figure 12.14 Measurements for liquid water at (a) 25 °C and (b) 75 °C, showing the value of the self-correlation function $G_s(r,t)$ as a function of t for various values of t (From Brockhouse et al.[16], by courtesy of I.A.E.A.)

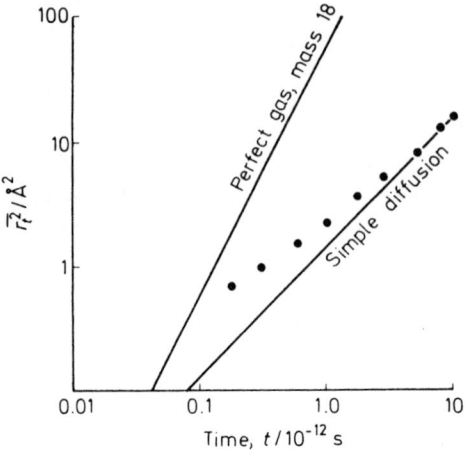

Figure 12.15 The variation of $\overline{r_t^2}$ with t for liquid water at 25 °C shown in comparison with computed curves for a gas (of mass 18) and a simple macroscopic model of diffusion (From Sakamoto et al.[17], by courtesy of J. Phys. Soc. Japan)

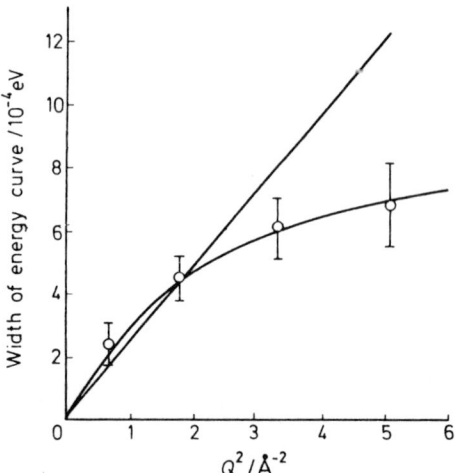

Figure 12.16 Variation of the line-width of the quasi-elastic scattering peak of water (as in Figure 12.13), at 293 K, with Q^2. The straight line, which holds at small values of Q, i.e. large values of r,t, corresponds to simple diffusion (From Larsson et al.[18], by courtesy of Ark. Fys. and Academic Press)

described adequately by the macroscopic diffusion equation when $t > 10^{-12}$ s. An alternative approach is to study the width of curves of quasi-elastic scattering, such as that of *Figure 12.13*, for various values of Q. For small values of Q, which correspond to large values of r,t, the width is proportional to Q^2, which is what would be expected on a model of simple diffusion. At larger values of Q the width becomes proportional to Q as expected. The experimental data of Larsson et al.[18]

are shown in *Figure 12.16*. For a detailed discussion of these studies of the diffusion of water molecules the reader is referred to an article by Powles[19].

An interesting application of the study of the inelastic scattering from water is found in an examination of cement pastes by Harris, Windsor and Lawrence[20]. This is one of the few reported uses of neutron scattering to solve practical problems and is based on the fact that the strongest

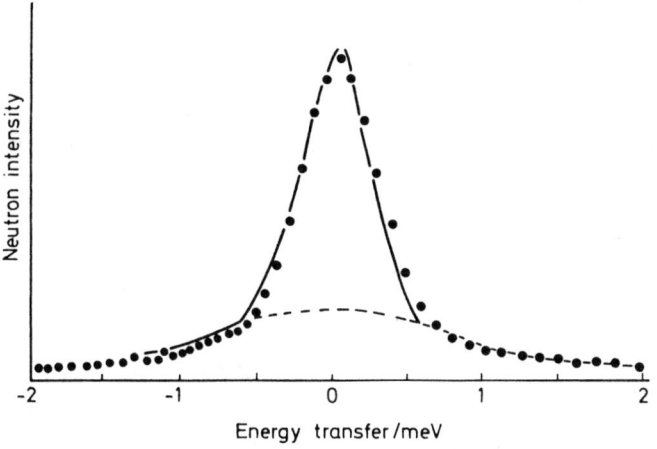

Figure 12.17 The 'quasi-elastic' scattering portion of the neutron scattering spectrum of a cement paste, as a function of the energy transfer in meV. The experimental points are shown in relation to a full-line curve which is compounded from a broad curve (dependent on diffusion of the free water) and a sharp curve convoluted with the experimental resolution curve (From Harris, Windsor and Lawrence[20], by courtesy of Magazine of Concrete Research)

cement products are those made with the least water, and also hydrated the most completely so that their free-water content is a minimum. In assessing the free-water content the usual method has been to measure the weight loss during drying, but there has always been uncertainty concerning the extent to which combined, hydrated, water is driven off at the same time. The inelastic scattering measurements are able to assess the free water quite accurately and established that its amount was substantially less than what was previously thought. The energy spectrum obtained with neutrons in the quasi-elastic scattering region is shown in *Figure 12.17* and consists very clearly of two components. These data were obtained at a scattering angle of 90° for incident neutrons of wavelength 4.26 Å and an energy width of 0.4 meV. The broad distribution is identical to the spectrum for pure water, and gives the amount of free water, whereas the sharp peak represents the combined water. In the particular case shown the free water represented a fraction of 0.21 of the weight of the sample compared with 0.30 as estimated from the loss on drying. Later experiments have studied the spectra at higher values of energy transfer and deduced the proportions of various compounds, such as $Ca(OH)_2$, which are present. The observed weightings have been correlated with strength and porosity.

12.2 Glasses

Glasses occupy an intermediate position between the long-range order of a crystalline solid and the purely short-range order, with diffusion of the atoms through the body of the material, which exists in a liquid. There is no long-range order and the diffraction pattern is diffuse, as illustrated in the upper curve of *Figure 12.18* (Hansen, Knudsen and Carneiro[21]), which shows a pattern for selenium glass. There are no sharp Bragg peaks and accordingly it is not possible to make a ready separation of the elastic scattering: the separation of elastic and inelastic scattering requires an energy analysis. The well-defined peaks in the pattern of polycrystalline selenium are contrasted in the lower portion of *Figure 12.18*. On the other hand, as in a solid, the atoms do have defined

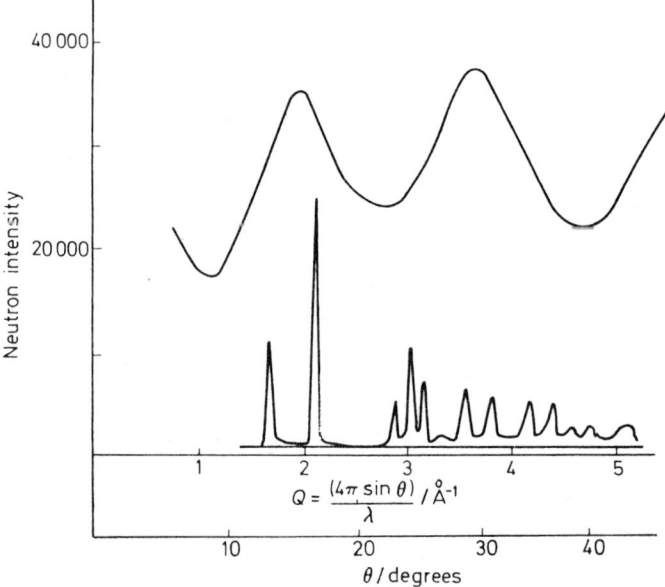

Figure 12.18 A comparison of the diffuse neutron diffraction pattern of selenium glass, for a wavelength of 1.70 Å, with the well-defined peaks of a pattern for polycrystalline selenium (From Hansen, Knudsen and Carneiro[21], by courtesy of J. Chem. Phys.)

equilibrium positions and there is a clear distinction between elastic and inelastic scattering, without the confusion posed by the quasi-elastic scattering of a liquid. In principle, therefore, it should be easier to determine the structure of a glass than a liquid, but a number of practical considerations invalidate such a conclusion. First, there is no range of simple monatomic glasses which can be studied, in the manner in which the liquid metals were so informative, and, apart from selenium and vapour-deposited germanium and silicon, all glasses are at least diatomic. Moreover, the simplest glasses, such as GeO_2 and SiO_2, do not contain isotopes of widely-varying scattering power which would permit the ready determination of component structure factors.

It is useful to consider separately the elastic and the total scattering from a glass. By considering that the time-averaged position of the atoms may be described in terms of static equilibrium positions and thermal displacements from these positions, it can be shown that the differential cross-section for elastic scattering is

$$\left(\frac{d\sigma}{d\Omega}\right)_{elastic} = N(\bar{b})^2 [1 + \int \rho g_e(r) \exp(iQ \cdot r) \, dr] \exp(-2W) \quad (12.21)$$

where $g_e(r)$ is the pair-distribution function for the equilibrium atomic positions and $\exp(-2W)$ is a Debye-Waller factor. This expression is strictly true only for a monatomic glass. In other cases separate values of $g(r)$ and W need to be used for each kind of atom. Lorch[22], measuring only the elastic scattering for vitreous silica with a triple-axis spectrometer and making correction for the Debye-Waller terms, has been able to deduce the form of $g_e(r)$. This is shown as the full-line curve in *Figure 12.19*.

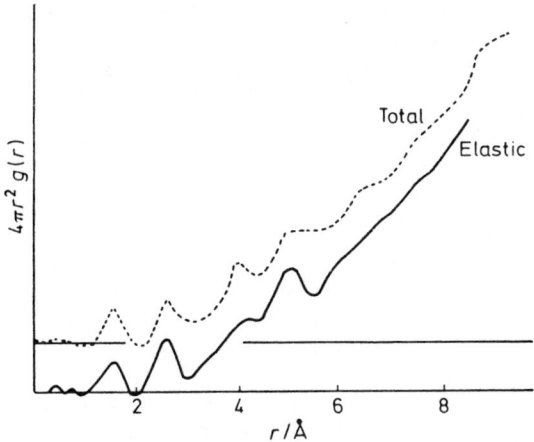

Figure 12.19 A comparison of the radial distribution functions $4\pi r^2 g(r)$ obtained from the elastic (curve ———) and the total (curve .....) scattering from vitreous silica (From Lorch[22], by courtesy of J. Phys. C)

On the other hand, the expression for the *total* scattering contains the factor κ/κ_0 and can only be evaluated precisely for X-rays, where the transfer of energy is negligibly small, thus giving the so-called 'static approximation' whereby

$$\left(\frac{d\sigma}{d\Omega}\right)_{total} = N(\bar{b})^2 [1 + \rho g(r) \exp(iQ \cdot r) \, dr] \quad (12.22)$$

For neutrons an approach to this approximation can be made by applying the Placzek corrections, which we discussed for liquids, and in this

170 Liquids, Glasses and Gases

way the second, dotted, curve in *Figure 12.19* was arrived at. There appears to be a significant difference between the two curves in the range of r between 3 and 4 Å.

It must be emphasized that these curves can only be approximate since no account has been taken of the three different distribution functions g_{Si-Si}, g_{O-O} and g_{Si-O}, nor of the different Debye-Waller factors for Si and O and it is a weighted mean value of $g(r)$ identified as $D(r)$ which is measured. Only very recently have the first experiments utilizing isotopic substitution to determine the individual functions been carried out with some titanium glasses. Previous to this, further interpretation has rested on comparing the measured diffraction patterns with the results of calculations for various model structures, assuming knowledge of the crystal chemistry of the material. A useful method of proceeding has been described by Leadbetter and Wright[23] who used a quasi-crystalline model in which the correlation function for a crystalline modification of the glass is amended by a function $F(r)$. This function is unity at low values of r and equal to zero at values of r greater than some value $r = L_0$ which is defined as the maximum correlation-length for the glass. Leadbetter and Wright[24] have used this procedure to study vitreous germania using both X-ray data and the neutron data of Lorch[25] and postulating several quasi-crystalline models based on the various known *crystalline* forms of GeO_2 which are analogous respectively to α-quartz, α-cristobalite and rutile. *Figure 12.20* (Leadbetter[26]) compares the neutron

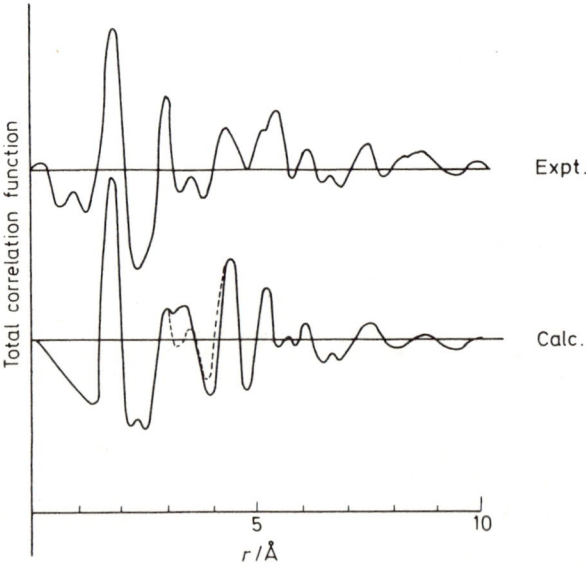

Figure 12.20 *A comparison of the neutron diffraction data for the total correlation function of vitreous germania with calculations according to two different models. The full line, in the lower section of the diagram, is for a model based on the structure of quartz with $L_0 = 10.5$ Å. The dotted line, for a model in which GeO_4 tetrahedra are rotated, gives improved agreement with experiment (From Leadbetter[26], by courtesy of Oxford University Press)*

data with calculations for a model which is based on the quartz structure with a value of L_0 = 10.5 Å. The agreement is very good except in the region between 2 and 4 Å. It was found in fact that in this region of r the X-ray data were in better agreement with the model calculations, suggesting therefore that the model was deficient in describing the oxygen positions, which would affect the neutron intensities more substantially. A revised model in which the structure was partially unwound by relative rotation of GeO_4 tetrahedra gave a substantial improvement when compared with the neutron data, as shown by the dotted curve in the figure. The effect of rotating the tetrahedra in this way is to increase the separation of some of the pairs of oxygen atoms.

It is interesting to compare the neutron data for GeO_2 in *Figure 12.20* with those obtained by Sinclair et al.[27] using a time-of-flight spectrometer installed at an electron linear accelerator. In this latter case a range of values of Q up to 35 Å^{-1} was included, resulting in a substantial improvement in the correlation function compared with Lorch's original data which only extended to Q = 18 Å^{-1}. The resulting curve for $D(r)$ is shown in *Figure 12.21* and is noteworthy for the resolution of the first Ge$\cdots$Ge distance which is clearly seen on the high-r side of the peak due to O$\cdots$O.

The *inelastic* scattering from a glass can be studied in a time-of-flight analysis but the interpretation is more difficult than for a crystal. For long wavelengths the vibrations in a glass will be elastic plane waves which

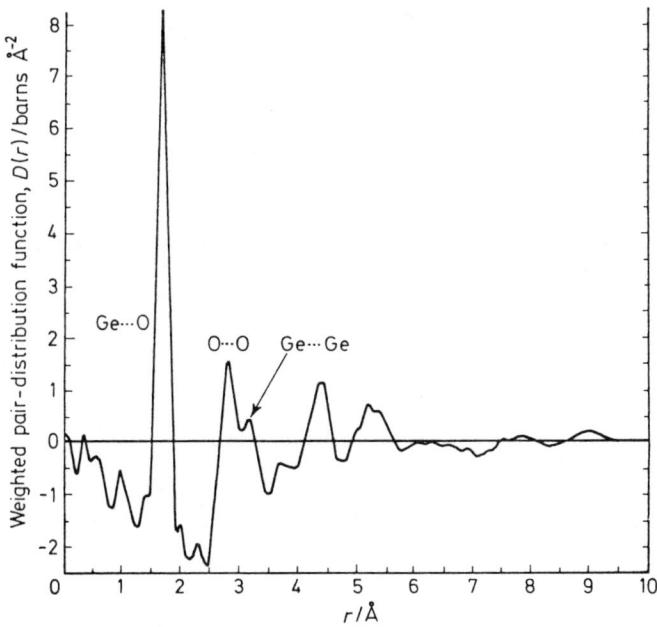

Figure 12.21 The weighted pair distribution function for vitreous GeO_2 measured by a time-of-flight spectrometer installed at a linear accelerator and giving values of Q up to 35 Å^{-1} (From Sinclair et al.[27], by courtesy of Nucl. Instrum. Methods)

can be described as phonons just as for crystals, although the phonon lifetimes will be much shorter corresponding to more heavily-damped waves, but at higher frequencies this approximation fails. Although most glasses are largely coherent scatterers, some useful information can be obtained by following the method of analysis described in Chapter 10 whereby the frequency spectrum of hydrogenous substances, which scatter incoherently, can be ascertained from measurements with polycrystalline materials. This follows because, as in equation 12.2 for liquids, the coherent cross-section is related to the total correlation function $G(r,t)$ which separates into the two parts $G_s(r,t)$ and $G_d(r,t)$ of equation 12.1. If experiments are made at high values of Q then the important part of G_d is that for low values of r: here the function will be very small, because in a glass (or any solid) only the *original* atom, not a different one, can be close to the origin.

Consequently when $Q \geqslant 2\pi/r_0$, where r_0 is the interatomic spacing, $G(r,t)$ will approximate to $G_s(r,t)$ and, from equations 12.4, 12.5, $S(Q,\omega)$ will approach $S_s(Q,\omega)$. Thus the form of the scattering pattern, which is $d^2\sigma_{coh}/d\Omega\, dE'$, will approach the form of $d^2\sigma_{incoh}/d\Omega\, dE'$ given in equation 10.9, except that the scattering amplitude $(\bar{b})^2$ will be relevant, rather than $(\overline{b^2}) - (\bar{b})^2$. Accordingly we arrive at the approximation

$$\frac{d^2\sigma}{d\Omega dE'} = N \frac{\kappa}{\kappa_0} \frac{Q^2}{2\omega} \left(\frac{1}{\exp(E/kT) - 1} + \frac{1}{2}(1 \pm 1) \right) Z(\omega)$$

$$\times \sum \frac{(\bar{b}_\nu)^2}{M} U_\nu^2 \exp(-2W_\nu) \qquad (12.23)$$

where the summation is over the ν atoms in a unit of glass of composition such as SiO_2 or GeO_2 and the disposable sign is chosen as $\pm$ according to whether the neutron loses or gains energy in the inelastic scattering process.

It is convenient to define a function $g(Q,\omega)$ which can be derived from the time-of-flight data in accordance with

$$g(Q,\omega) = \frac{d^2}{d\Omega dE'} \frac{\kappa_0}{\kappa} \frac{2\omega}{NQ^2} \exp(E/kT - 1) \qquad (12.24)$$

It follows by comparing equations 12.23 and 12.24 that

$$g(Q,\omega) = Z(\omega) \left[\left\{ \frac{b^2}{M} U_\omega^2 \exp(-2W) \right\}_{Ge} + 2 \left\{ \frac{b^2}{M} U_\omega^2 \exp(-2W) \right\}_O \right]$$

$$(12.25)$$

under conditions of 'energy gain' for the particular case of GeO_2. *Figure 12.22* shows the results of Leadbetter and Litchinsky[28] for vitreous GeO_2,

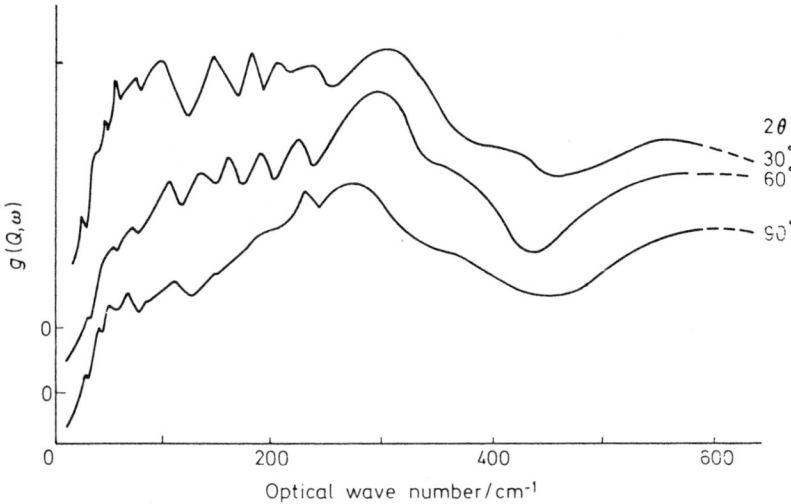

Figure 12.22 Curves of $g(Q,\omega)$ for vitreous silica measured at scattering angles of $2\theta = 30, 60$ and $90°$. The ordinates of the curves are displaced to avoid overlap (From Leadbetter and Litchinsky[28], by courtesy of Discuss. Faraday Soc.)

using time-of-flight data collected at three different angles of scattering for cold neutrons from which only energy-gain spectra would be seen. For a given value of ω, Q will increase with the scattering angle (for a gain of energy) so that a large value of scattering angle will approximate best to the assumption of incoherence which has been made above. In the same way, as ω increases Q will increase so that the results for different scattering angles tend to become the same. For values of ω between 200 and 600 cm^{-1} the curve for $g(Q,\omega)$ may be considered to represent $Z(\omega)$ which describes the frequency distribution in the solid.

A second example, of the way in which the dominance of $G_s(r,t)$ over $G_d(r,t)$ at large values of Q means that purely coherent scatterers will indicate the frequency distribution, is the comparison of α-quartz and vitreous silica by Leadbetter and Stringfellow[29]. These measurements were made with monochromatic neutrons whose wavelength could be varied from about 0.65 to 2.4 Å, covering an energy range from 0.015 to 0.2 eV, and which were detected after scattering by a counter covered by a beryllium filter. As the incident wavelength is varied, the counter assesses the number of neutrons which have suffered an energy loss ranging from about 1400 down to 120 cm^{-1}. The data obtained for vitreous silica are given in *Figure 12.23*, showing three main peaks close to 360, 800 and 1200 cm^{-1}. Although there are differences in detail this spectrum is broadly similar to that found for crystalline α-quartz (also shown in the figure). Evidently the main features depend largely on the properties of the SiO_4 tetrahedra but with smearing-out of the spectra because of the loss of long-range order in the glass.

For a detailed review of the use of neutron scattering for studying glasses the reader is referred to an article by Leadbetter[26].

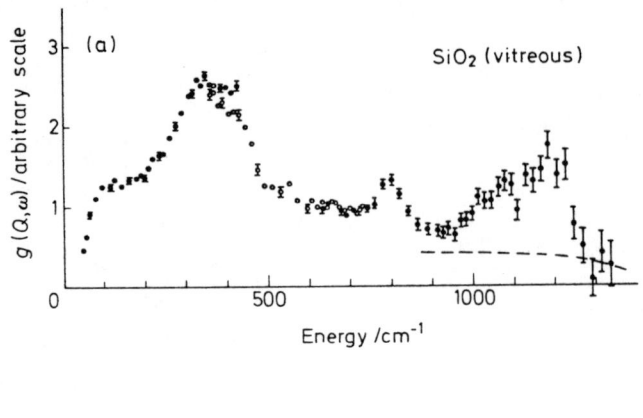

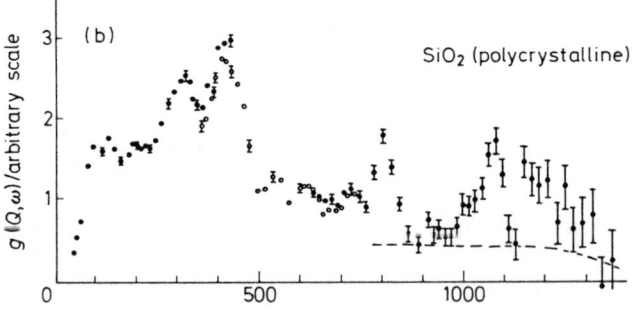

Figure 12.23 A comparison of curves of g(Q,ω) for (a) vitreous and (b) polycrystalline silica. The open and closed circles distinguish energy regions explored by using different monochromator planes (From Leadbetter and Stringfellow[29], by courtesy of I.A.E.A.)

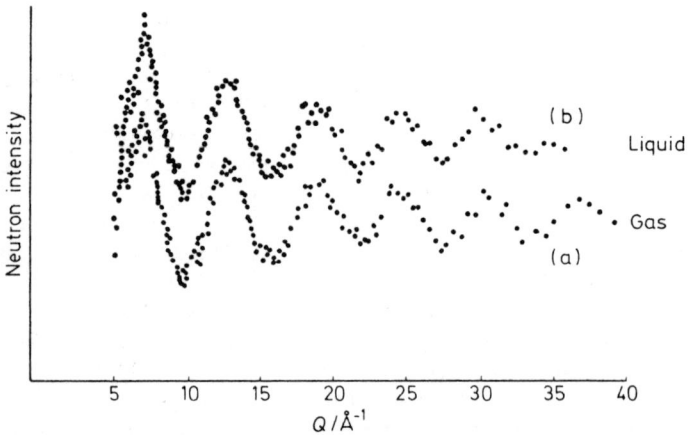

Figure 12.24 A comparison of the neutron scattering spectra of (a) gaseous nitrogen at 35 atmospheres pressure and (b) liquid nitrogen at 77 K; note that the experimental points extend to a Q value of 40 Å^{-1} (From Page and Powles[33], by courtesy of Mol. Phys.)

12.3 Gases

Very few measurements of the diffraction of neutrons by gases have been made since the pioneer work of Hurst and his collaborators[30-32] at Chalk River about 1950. We shall mention, however, a recent study of gaseous nitrogen by Page and Powles[33], because of the light which it throws on the determination of the structure of *liquids*. Measurements with nitrogen gas enclosed in a vanadium tube under a pressure of 35 atmospheres yielded a scattering pattern shown at (a) in *Figure 12.24*. A period of four days was expended, using the time-of-flight technique at a linear accelerator, in order to obtain a pattern having significant accuracy. The figure also shows at curve (b) a pattern for *liquid* nitrogen recorded with the same spectrometer and it is concluded that the oscillations, as a function of Q, are almost indistinguishable for liquid and gas. At large values of Q the period of these oscillations will depend only on the N–N separation within the molecule and it is concluded therefore that this separation is the same, to within ± 0.01 Å, for both liquid and gas. Accordingly it seems likely that the separation of 1.102 Å determined[34] by electron diffraction will also hold for the liquid, in disagreement with a previous conclusion from neutron work[35] that the distance in the liquid was 1.06 ± 0.01 Å. It has been shown by Page and Powles that the earlier neutron work was invalidated by the inadequacy of the Placzek corrections. They have developed much more adequate corrections which are generally applicable to the study of liquids containing relatively light nuclei.

References

1. VAN HOVE, L., *Phys. Rev.*, **95**, 249 (1954)
2. PLACZEK, G., *Phys. Rev.*, **86**, 377 (1952)
3. NORTH, D.M., ENDERBY, J.E. and EGELSTAFF, P.A., *J. Phys. C*, **1**, 784 (1968)
4. NORTH, D.M., ENDERBY, J.E. and EGELSTAFF, P.A., *J. Phys. C*, **1**, 1075 (1968)
5. PAGE, D.I., EGELSTAFF, P.A., ENDERBY, J.E. and WINGFIELD, B.R., *Phys. Lett.*, **A29**, 296 (1969)
6. ENDERBY, J.E., NORTH, D.M. and EGELSTAFF, P.A., *Phil. Mag.*, **14**, 961 (1966)
7. PAGE, D.I. and MIKA, K., *J. Phys. C*, **4**, 3034 (1971)
8. POWLES, J.G., *J. Phys. C*, **8**, 895 (1975)
9. NARTEN, A.H., VASLOW, F. and LEVY, H.A., *J. Chem. Phys.*, **58**, 5017 (1973)
10. ENDERBY, J.E., HOWELLS, W.S. and HOWE, R.A., *Chem. Phys. Lett.*, **21**, 109 (1973)
11. HOWE, R.A., HOWELLS, W.S. and ENDERBY, J.E., *J. Phys. C*, **7**, L111 (1974)
12. NEILSON, G.W., HOWE, R.A. and ENDERBY, J.E., *Chem. Phys. Lett.*, **33**, 284 (1975)

13. PAGE, D.I. and POWLES, J.G., *Mol. Phys.*, **21**, 901 (1971)
14. EGELSTAFF, P.A., PAGE, D.I. and POWLES, J.G., *Mol. Phys.*, **20**, 881 (1971)
15. DORE, J.C., CLARKE, J.H. and WENZEL, J.T., *Nucl. Instrum. Methods*, **138**, 317 (1976)
16. BROCKHOUSE, B.N., BERGSMA, J., DASANNACHARYA, B.A. and POPE, N.K., in *Inelastic Scattering of Neutrons in Solids and Liquids*, Vol.I, 189, I.A.E.A., Vienna (1963)
17. SAKAMOTO, M., BROCKHOUSE, B.N., JOHNSON, R.G. and POPE, N.K., *J. Phys. Soc. Japan*, **17**, Suppl. BII, 370 (1962)
18. LARSSON, K.E., DAHLBORG, U. and HOLMRYD, S., *Ark. Fys.*, **17**, 369 (1960)
19. POWLES, J.G., in *Chemical Applications of Thermal Neutron Scattering*, ed. B.T.M. Willis, 118, Clarendon Press, Oxford (1973)
20. HARRIS, D.H.C., WINDSOR, C.G. and LAWRENCE, C.D., *Magazine of Concrete Research*, **26**, 65 (1974)
21. HANSEN, F.Y., KNUDSEN, T.S. and CARNEIRO, K., *J. Chem. Phys.*, **62**, 1556 (1975)
22. LORCH, E., *J. Phys. C*, **3**, 1314 (1970)
23. LEADBETTER, A.J. and WRIGHT, A.C., *J. Non-cryst. Solids*, **7**, 23 (1972)
24. LEADBETTER, A.J. and WRIGHT, A.C., *J. Non-cryst. Solids*, **7**, 37 (1972)
25. LORCH, E., *J. Phys. C*, **2**, 229 (1969)
26. LEADBETTER, A.J., in *Chemical Applications of Thermal Neutron Scattering*, ed. B.T.M. Willis, 146, Clarendon Press, Oxford (1973)
27. SINCLAIR, R.N., JOHNSON, D.A.G., DORE, J.C., CLARKE, J.H. and WRIGHT, A.C., *Nucl. Instrum. Methods*, **117**, 445 (1974)
28. LEADBETTER, A.J. and LITCHINSKY, D., *Discuss. Faraday Soc.*, **50**, 62 (1970)
29. LEADBETTER, A.J. and STRINGFELLOW, M., in *Neutron Inelastic Scattering*, 501, I.A.E.A., Vienna (1973)
30. ALCOCK, N.Z. and HURST, D.G., *Phys. Rev.*, **75**, 1609 (1949)
31. ALCOCK, N.Z. and HURST, D.G., *Phys. Rev.*, **83**, 1100 (1951)
32. HURST, D.G. and ALCOCK, N.Z., *Can. J. Phys.*, **29**, 36 (1951)
33. PAGE, D.I. and POWLES, J.G., *Mol. Phys.*, **29**, 1287 (1975)
34. BARTELL, L.S. and KUCHITSU, K., *J. Phys. Soc. Japan*, **17**, Suppl.BII, 20 (1962)
35. POWLES, J.G., *Mol. Phys.*, **26**, 1325 (1973)

APPENDIX 1

Coherent scattering amplitudes of elements and isotopes for neutrons* (10^{-12} cm)

Element	Isotope	b	Element	Isotope	b
H	^{1}H	−0.374	Fe		0.95
	^{2}H	0.667		^{54}Fe	0.42
	^{3}H	0.47		^{56}Fe	1.01
	^{4}H	0.30		^{57}Fe	0.23
He	^{4}He	0.30	Co	^{59}Co	0.28
Li		−0.214	Ni		1.03
	^{6}Li	0.18 + 0.025 i		^{58}Ni	1.44
	^{7}Li	−0.233		^{60}Ni	0.28
Be		0.774		^{61}Ni	0.76
B		0.54 + 0.021 i		^{62}Ni	−0.87
	^{11}B	0.60		^{64}Ni	−0.04
C		0.665	Cu		0.76
	^{12}C	0.665		^{63}Cu	0.67
	^{13}C	0.60		^{65}Cu	1.11
N	^{14}N	0.94			
	^{15}N	0.65	Zn		0.57
O	^{16}O	0.580		^{64}Zn	0.55
	^{17}O	0.578		^{66}Zn	0.63
	^{18}O	0.600		^{68}Zn	0.67
F	^{19}F	0.56			0.72
Ne		0.46	Ga		0.82
Na	^{23}Na	0.36	Ge		0.64
Mg		0.52	As		0.80
	^{24}Mg	0.55	Se		0.68
	^{25}Mg	0.36	Br		0.74
	^{26}Mg	0.49	Kr		0.71
Al	^{27}Al	0.35	Rb	^{85}Rb	0.83
Si		0.42			0.69
P		0.51	Sr	^{89}Y	0.76
S		0.28	Y		0.71
Cl		0.96	Zr		0.71
	^{35}Cl	1.18	Nb		0.69
	^{37}Cl	0.26	Mo		0.68
A		0.20	Tc		0.73
	^{36}A	2.43	Ru		0.58
K		0.37	Rh		0.60
	^{39}K	0.37	Pd		0.60
Ca		0.47	Ag	^{107}Ag	0.83
	^{40}Ca	0.49		^{109}Ag	0.43
	^{44}Ca	0.18			0.37 + 0.16 i
Sc	^{45}Sc	1.18	Cd	^{113}Cd	−1.5 + 1.2 i
Ti		−0.34			0.39
	^{46}Ti	0.48	In		0.62
	^{47}Ti	0.33	Sn	^{116}Sn	0.58
	^{48}Ti	−0.58		^{117}Sn	0.64
	^{49}Ti	0.08		^{118}Sn	0.58
	^{50}Ti	0.55		^{119}Sn	0.60
V	^{51}V	−0.038		^{120}Sn	0.64
Cr		0.352		^{122}Sn	0.55
	^{52}Cr	0.490		^{124}Sn	0.59
Mn	^{55}Mn	−0.37	Sb		0.56

Appendix 1

Element	Isotope	b	Element	Isotope	b
Te		0.58	Lu		0.73
	^{120}Te	0.52	Hf		0.78
	^{123}Te	0.57	Ta		0.70
	^{124}Te	0.55	W		0.48
	^{125}Te	0.56		^{182}W	0.83
I	^{127}I	0.53		^{183}W	0.43
Xe		0.48		^{184}W	0.76
Cs		0.55		^{186}W	−0.12
Ba		0.52	Re		0.92
La		0.83	Os		1.07
Ce		0.48		^{188}Os	0.78
	^{140}Ce	0.47		^{189}Os	1.10
	^{142}Ce	0.45		^{190}Os	1.14
Pr	^{141}Pr	0.44		^{192}Os	1.19
Nd		0.77	Ir		1.06
	^{142}Nd	0.77	Pt		0.95
	^{144}Nd	0.28	Au		0.76
	^{146}Nd	0.87	Hg		1.27
Pm			Tl		0.89
Sm			Pb		0.94
	^{149}Sm	−1.9 + 4.5 i	Bi		0.86
	^{152}Sm	−0.5	Po		
	^{154}Sm	0.96	At		
Eu		0.68	Rn		
Gd		1.5	Fr		
	^{157}Gd	4.3 + 4 i	Ra		
	^{160}Gd	0.91	Ac		
Tb	^{159}Tb	0.76	Th	^{232}Th	1.03
Dy		1.69	Pa		1.30
	^{160}Dy	0.67	U		0.85
	^{161}Dy	1.03		^{235}U	0.98
	^{162}Dy	−0.14		^{238}U	0.85
	^{163}Dy	0.50	Np		1.05
	^{164}Dy	4.94	Pu		0.75
Ho	^{165}Ho	0.85		^{240}Pu	0.35
Er		0.79		^{242}Pu	0.81
Tm		0.72	Am	^{243}Am	0.76
Yb		1.26	Cm	^{244}Cm	0.7

*For complex scattering amplitudes the values are given for $\lambda = 1$ Å

APPENDIX 2

Elements and isotopes showing significant incoherent scattering

Element or isotope	Cross-section/10^{-24} cm^2		Element or isotope	Cross-section/10^{-24} cm^2	
	S coherent	s incoherent		S coherent	s incoherent
H	1.8	79.7	As	5.1	2.9
D	5.6	2.0	Sr	6.0	4.0
Li	0.58	0.7	Rh	4.4	1.2
^{7}Li	0.68	0.8	Ag	4.5	1.8
Na	1.65	1.7	^{109}Ag	2.3	3.7
Cl	11.5	3.5	Te	3.7	0.8
A	0.5	0.4	Cs	3.7	3.2
K	1.72	0.5	Ba	3.5	2.5
Sc	17.5	6.0	Pr	2.4	1.6
Ti	1.4	3.0	Nd	6.5	9.5
V	0.03	5.1	Ho	9.1	4
Cr	1.57	2.5	Er	7.8	7
Co	0.79	5.2	W	2.9	2.8
Ni	13.3	4.7	Au	7.3	2
Cu	7.3	1.2	Hg	20.1	6
Ga	6.5	1.0			

APPENDIX 3

Table of corresponding values of wavelength (Å) and energy (eV) for neutrons, together with equivalent values of optical wave-number (cm^{-1}) for energy transfer

λ/Å	Energy/eV	Optical wave-number/cm^{-1}
25	0.00013	1.05
20	0.00024	1.65
15	0.00036	2.93
11.5	0.00062	**5**
10	0.00082	6.59
9.04	**0.001**	8.06
9	0.00101	8.13
8.12	0.00124	**10**
8	0.00128	10.3
7	0.00167	13.4
6	0.00227	18.3
5.74	0.00248	**20**
5.0	0.00327	26.4
4.01	**0.005**	41.0
4.0	0.00511	41.2
3.63	0.00620	**50**
3.0	0.00908	73.2
2.86	**0.01**	80.5
2.57	0.0124	**100**
2.5	0.0131	105
2.0	0.0204	165
1.82	0.0247	**200**
1.5	0.0363	293
1.28	**0.05**	402
1.15	0.0618	**500**
1.0	0.0818	659
0.904	**0.1**	806
0.9	0.101	813
0.812	0.124	**1000**
0.8	0.128	1030
0.7	0.167	1340
0.663	0.186	**1500**
0.640	**0.2**	1610
0.6	0.227	1830
0.574	0.248	**2000**
0.5	0.327	2640

BIBLIOGRAPHY

1. BACON, G.E., *Neutron Diffraction*, 3rd edn., Clarendon Press, Oxford (1975)
2. BACON, G.E., *Neutron Physics*, Wykeham Press, London (1969)
3. BOUTIN, H. and YIP, S., *Molecular Spectroscopy with Neutrons*, M.I.T. Press, Cambridge, Mass., U.S.A. (1968)
4. EGELSTAFF, P.A. (ed.), *Thermal Neutron Scattering*, Academic Press, London (1965)
5. GUREVICH, I.I. and TARASOV, L.V., *Low-Energy Neutron Physics*, North Holland, Amsterdam (1968)
6. IZYUMOV, Yu. A. and OZEROV, R.P., *Magnetic Neutron Diffraction*, Plenum Press, New York (1970)
7. LAROSE, A. and VANDERWAL, J. (eds), *Scattering of Thermal Neutrons: A Bibliography (1932-74)*, Plenum Press, New York (1974)
8. MARSHALL, W. and LOVESEY, S.W., *Theory of Thermal Neutron Scattering*, Clarendon Press, Oxford (1971)
9. OLES, A., KAJZAR, F., KUCAB, M. and SIKORA, W. (eds), *Magnetic Structures Determined by Neutron Diffraction*, Polska Akademia Nauk, Warsaw (1976)
10. WILLIS, B.T.M. (ed.), *Thermal Neutron Diffraction*, Clarendon Press, Oxford (1970)
11. WILLIS, B.T.M. (ed.), *Chemical Applications of Thermal Neutron Scattering*, Clarendon Press, Oxford (1973)
12. *Neutron Inelastic Scattering 1972:* Proceedings of a Symposium held at Grenoble, France; International Atomic Energy Agency, Vienna (1972)

INDEX

Absorption, 25, 150
Acetic acid, 128
Acetylene, 58
L-Alanine, 63
Alloys, liquid, 156
Amino-acid derivatives, 62
Ammonium
 hydrogen phosphate, 36
 perchlorate, 125
 sulphate, 125
Annulene, 48
Anomalous scattering, 7, 83
Aqueous solutions, 158
L-Arginine, 63
Argon, 155
L-Asparagine, 63

Barium
 chloride solution, 160
 tungstate, 51
Barn, 8
Beryllium-filter
 detector, 27, 120
 spectrometer, 30
Biological materials, 82
Bond lengths, 40
Bond shortening, 39, 41, 65
Bound/free cross-section, 10
Breit-Wigner formula, 7

Cadmium-113, 83
Cadmium nitrate, 83
Caesium hydrogen chloride, 121
Calcium
 fluoride, 107
 molybdate, 57
Catalysis, 128
Cement, 167
Chromium(III)
 dioxygen hydride, 45, 124
 fluoride, 102
 ion, 102

Clay minerals, 133
Clusters, 106, 109
Cobalt
 carbonyl hydrides, 129
 dioxygen hydride, 45, 125
Coherent nuclear scattering, 17, 138, 152
 table of, 177
'Cold' source, 20, 31
Collimation, 20
Complex scattering amplitude, 83, 177
Compound nucleus, 9
'Constant-Q' method, 26
Copper, 25
 chloride, 156
 –tin alloys, 156
Correlation functions, 117, 151
Counters, 22, 89
Covalency, 92
 resonance methods for, 100
Crystal
 monochromators, 15, 22
 vibrations, 4, 118, 138
Cyanuric acid, 75
Cyclodecane-1,6-diol, 47

Detectors, 22, 89
Deuterium
 bromide, 53
 chloride, 53
 solid, 58
 substitution, 11, 36, 82, 86, 114, 119
 selective, 127
 sulphide, 53
Diffraction, 1, 104
Diffractometers, 21, 25
Diffuse scattering, 4, 104, 109
Diffusion, 133, 146, 150, 164
'Direct' methods, 69
Dispersion law, 118, 138

Elastic scattering, 4, 20, 116
Electron diffraction, 1
Energy of neutron, 20, 116, 180
  measurement of, 21
Energy-gain method, 31, 123
Energy-loss method, 27, 123
Energy-momentum relation, 20, 116
Energy transfer, 20, 116, 180
Experimental methods, 19, 87

Ferritin, 87
Ferro-electrics, 36
Fick's Law, 164
Forbidden reflections, 97
Form-factor, 6, 12, 61, 93
Fourier analysis, 2, 154

Gases, 175
Germanium(IV) oxide, 170
Glasses, 168
Glycollic acid, 70
Guide-tubes, 32, 87

Haemoglobin, 87
Heavy-element compounds, 48, 61, 105
Hexamethylenetetramine, 74, 118
High-temperature, measurements at, 105, 150
'Hot' source, 20, 31, 126
Hydrated compounds, 41
Hydrides, 113, 126, 129
Hydrogen
  atom, location, 39, 69
  bond, 41, 120
  chloride, 53
  /deuterium, comparison, 9
   substitution, 11, 36, 82, 86, 114, 119
  incoherent scattering of, 9, 119

Ice, 42
Incoherent scattering, 18, 116, 138
  table of, 179
Inelastic scattering, 4, 20, 116, 171
Insulin, 84
Ionic solutions, 134
Iron carbonyl dihydride, 129

Iron(III) ion, 95
Iron(II) oxide, 112
Isotope
  incoherence, 8, 104
  substitution, 156, 170
    see also hydrogen/deuterium substitution

Lanthanum
  chromate, 102
  ferrate, 95
Lead, 150
Light elements, 11
Linear accelerator, 20, 171, 175
Liquids, 150
Liquid metals, 154
Lithium
  chloride, 134, 158
  formate, 80
Lone pairs, 75
Long-wavelength neutrons, 31, 32, 89
Low temperature, measurements at, 36, 53
L-Lysine, 65

Macromolecules, 85
Magnetic interaction vector, 14
Magnetic materials, 12
Magnetic moment
  of ions, 92, 112
  of neutrons, 1
Magnetic scattering, 3, 6, 11, 18, 92
Magnetic structures, 13, 52
Magnetic unit-cell, 13
Manganese
  carbonate, 98
  carbonyl trihydride, 130
  fluoride, 96, 102
  oxide, 102
  selenide, 102
  sulphide, 102
  telluride, 102
Manganese(II) ion, 102
Manganese(IV) ion, 102
MARX spectrometer, 26
Material, choice of, 32
Melampodin, 71
Mercury chromate, 48

Metals, liquid, 154
Methane, 131
Methanol, 127
Methyl methacrylate, 144
Methyl siloxane, 146
Miller indices, 3
Molecular rotation, 47, 65, 133, 135, 142
Molecular spectroscopy, 116
Momentum of neutron, 20, 116
Momentum transfer vector, 20
Monochromator, 22
Montmorillonite, 133
Multi-counters, 22, 89
Muscle, 89
Myoglobin, 82

Neutron
  magnetic moment of, 1
  spectrum of, 22
  wavelength of, 19
Nickel, 22
  chloride solution, 160
  oxide, 95
Nickel(II) ion, 95, 100
Niobium carbide, 113
Nitrogen, 175
Nuclear reactor, 20
Nuclear scattering, 3, 6, 17
  table of, 177
Nuclear size, 6

Overcrowding, 45
Oxalic acid dihydrate, 77
Oxonium ion, 43
Oxygen, solid, 60

Pair-distribution function, 154
Partial structure-factor, 156
Patterson function, 69
Pauli spin-operator, 16
Perovskites, 36
Perspex, 144
Phenanthrene, 45
Phonons, 4, 26, 118, 138, 171
Phosphonium iodide, 125
Placzek corrections, 154, 169, 175
Polarization analysis, 9, 16, 163
Polarized neutrons, 9, 14, 97
Polarizing crystal, 15, 16

Polyethylene, 138
  oxide, 149
Polymers, 138
Poly(propylene oxide), 149
Polytetrafluoroethylene, 138, 143
Potassium
  hydrogen bisphenylacetate, 42
  hydrogen bistrichloroacetate, 125
  hydrogen chloromaleate, 70
  hydrogen fluoride, 42, 120
  iodide, 134
  nickel fluoride, 95
  rhenium hydride, 126
  trihydrogen disuccinate, 70
Position-sensitive detectors, 26, 89
Potential scattering, 7
Powder measurements, 22, 33, 118
Pressure, measurements under, 150
Profile refinement, 33, 52
L-Proline, 70
Pulsed reactor, 20

Quartz, 173
Quasi-elastic scattering, 128, 133, 146, 151, 166
Quasi lattice, 161

Radius of gyration, 84
Refinement
  joint of X-ray and neutron data, 73
  powder profile, 33, 52
Reliability factor, 34
Resonance methods in covalency, 100
Resonance scattering, 7
Rhenium carbonyl trihydride, 130
'Riding' motion, 41
Rigid-body motion, 47, 65
Rubidium, 155

Samarium, 83
Sayre's equation, 69
Scattering
  diffuse, 4
  elastic/inelastic, 4, 20, 116, 171
  incoherent, 8, 9, 18, 116, 138, 152, 179

Scattering *continued*
  magnetic, 3, 6, 11, 18, 92
  nuclear, 3, 6, 17
  quasi-elastic, 128, 133, 146, 151, 166
  small-angle, 31, 84
Scattering amplitude, 3, 6, 12, 177
  cross-section, 8, 10, 152
  law, 117, 153
  length, 6, 12, 83
    complex, 7, 83, 177
  vector, 14
Secondary extinction, 28, 39
Selection rules, 119
Selenium glass, 168
L-Serine, 71
Shielding, 25
Silica, 169
Single-crystal measurements, 24, 33, 118
Size of crystal, 39, 49
Small-angle scattering, 31, 84
Sodium
  chloride solution, 160
  hydrogen fluoride, 42
  rhenium hydride, 126
Solutions, 134, 158
Space-time correlation functions, 117, 151
Spectrometers, 21, 25, 26, 30
Spectrum of neutrons, 22
Spin-flip device, 16
Spin incoherence, 8, 9, 104
Spin polarization, 102
Static approximation, 154, 169
Strontium
  molybdate, 51
  tungstate, 51
Structural analysis, 39
Structure factor of liquid, 154, 162
Sucrose, 39
Sulphosalicylic acid, 42
Surface chemistry, 128

Techniques, experimental, 19, 87
Teflon, 143
Tetracyanoethylene oxide, 77

Tetramethylammonium bromide, 135
Thermal motion, 3, 41, 55, 62, 73, 77, 164
Thermal parameters, 39
Time-of-flight technique, 28, 171
Titanium diniobate, 36
Toluidinium bifluoride, 43
Triazine, 76
Trichloroacetic acid, 74
Trichlorobenzene, 118

Unit cell, 1, 104
  magnetic, 13
Uranates, 34
Uranium
  arsenide, 52
  compounds, 52
  diantimonide, 52
  oxides, 49, 105
  selenide, 52

Vanadium carbide, 112
Vermiculite, 133
Vibrations, 4
  distribution function of, 118, 138
Vitamin B_{12}, 82

Water, 150, 162
Wavelength of neutron, 19, 180
'White' neutrons, 28

Xenon dioxide difluoride, 61
X-ray/neutron
  comparison of bond lengths, 65, 80
  co-ordinate differences, 73
  syntheses, 73

Yttrium
  ferrate, 95
  oxalate, 45

Zeolites, 130
Zwitterion, 63

/541.28B128N>C1/